M000237987

Tornados, Rattlesnakes & Oil

Tornados, Rattlesnakes & Oil

A Wildcatter's Memories of Hunting for "Black Gold"

Thomas E. Cochrane, Geologist

River Beach Press
The Sea Ranch, CA

RIVER BEACH PRESS
Tornados, Rattlesnakes & Oil — **A Wildcatter's Memories** of Hunting for "Black Gold"

Copyright © 2018 by Thomas E. Cochrane
All Rights Reserved

Book development, editor, cover/interior design, and promotion: S.A. Jernigan, Renaissance Consultations
(www.MarketingAndPR.com)

This book was self-published by Thomas E. Cochrane under River Beach Press, all photographs are the property of Thomas E. Cochrane. No part of this book may be reproduced in any form or by any electronic or mechanical means, including information storage and retrieval systems, without written permission, except in the case of a reviewer who may quote brief passages embodied in critical articles or in a review.

If you would like to do any of the above or purchase individual or bulk copies, please contact:

Cochrane Enterprises, LLC, PO Box 358, The Sea Ranch, CA 95497, www. RiverBeachPress.com

Published in the United States by River Beach Press
Printed in the United States
ISBN: 978-0-9985106-1-3
Library of Congress Control Number: 2018901113
Subjects: 1.) Biography/Science   2.) Biography/Business   3.) Biography/Historical

Trademarked names appear throughout this book. Rather than use a trademark symbol with every occurrence of a trademarked name, names are used in an editorial fashion, with no intention of infringement of the respective owner's trademark.

The publisher believes, and we acknowledge others strongly disagree, that memoir allows the writer to work from memory instead of from a strict journalistic or historical standard. The firsthand recollections herein are being published in 2018 and relate primarily to the period of time from 1964 - 1986. The information provided within this book is for general informational purposes only. While we have attempted to provide up-to-date and correct information, there are no representations or warranties, express or implied, about the completeness, accuracy, reliability, suitability, or availability with respect to the information contained within.

Dedication

    This book is dedicated to all the people I met in the oil patch over the years, for what I learned from them, and for what they taught me to do with that knowledge. Time and distance have blurred my memories and some of their names have escaped me, or possibly I attributed a certain story to the wrong person. In my mind, the facts, stories, and opinions are as true as I remember them. However, stories tend to get better with time, and I am sure my love, Susan, and some of our friends think that I have exaggerated some of these incidents.

◆ The Chapters of My Life — both in & out of the oil patch ◆

Preface

If you knew me as a child, you would never have predicted I would end up in the oil business, which has been and is currently perceived as highly destructive to nature and the environment. How and why did we take the paths which lead us to the present?

I grew up outside in the natural world: rain or shine, snow or sleet. When school was out, I was outside, period. We lived on the outskirts of a small town in upstate New York. My grandfather owned the neighboring farm. Milking cows, doing farm activities, watching animal births, killing chickens, and the slaughtering of hogs were all part of farm life. My uncle and my best friend's dad were both hunters and fishermen. By age 10, I was spending many hours with my uncle fishing and hunting woodchucks. Dave's father taught us how to tie flies and fly fish. Later, he taught us to shoot skeet. By age 16, I finally had a hunting license.

Hunting and fishing are quiet activities where one spends much of the time just watching and waiting, and of course thinking about nature. Many times, without even firing a shot, I just quietly studied the deer, the squirrels, the rabbits, and the birds. I finally gave up hunting with a gun by age 30. Since then, I hunt wildlife with a camera instead.

Essentially I was a loner, although I usually had one best friend with whom I spent lots of time. I didn't join in group sports or organized activities for the most part. My interests were in nature…and I had lots of questions. At bedtime, I spent many hours into the night reading nature writers like Jack London, John Muir, Henry Thoreau, Aldo Leopold, and others. There were also the Hardy Boys books, the Tarzan books, the travel books—like Richard Halliburton, I too wanted to see the world, especially the natural world.

I knew from an early age I would be a scientist, but I didn't know what kind. By my teenage years, I was collecting fossils and rocks. The glacial outwash gravels found in and along our rivers and streams contained rounded pebbles of quartzite, granite, and other kinds of rock which were different from the local bedrock. In the whole southern tier of New York State's counties, there were mostly just fine-grained sandstones and shale of Devonian age. The glacial gravels must have come from the Adirondack Mountains in northern New York State, or even from Canada.

My love of the outdoors obviously drew me into geology—but I didn't declare that as my major until I was a junior in college, although given my years of enthusiastic rock-hounding, I'm sure my mother and sister knew I was to become a geologist!

Admittedly, economic pressures often draw most of us into professions or businesses. I had no knowledge of the oil business before I became a part of it. There were no oil or gas wells around my part of New York State, although the petroleum business put out lots of good PR in advertisements to promote their image. During my career in the oil business, whenever I encountered a problem I tried to correct it, or report it to someone who could handle it properly. However I'm not sure that was always the case inside the industry as a whole.

As with anything I became interested in during my lifetime, I got into it all the way and had a good time doing it. This book tells the tale of the oil industry which shaped my life and led me down the many paths I'll share with you, my reader, in the following pages.

COVER PHOTO: Blowout of gas well in Canton, Oklahoma -- incident occurred August 14, 1964 and burned until June 10, 1965. The drilling rig was completely destroyed. Two relief wells were drilled to intersect the wellbore. The well in the picture is only 200' away from the burning well. The first relief well was produced at maximum production to reduce the pressure. Over the course of nearly a year, six tries were made to kill the well. In the second relief well, 40,000 barrels of water finally succeeded in extinguishing the fire.

I took this photo two to three months after the blowout occurred.

Acknowledgements

The oil and gas industry collects many characters—maybe more so than some other industries—and, in this writing, a few are introduced to you. It takes many people to find and develop an oil or gas field. The industry slowly changed over time, but many of the descriptions of oil rigs, company philosophies, leasing, and money-raising are still pertinent in today's oil patch.

My time in the industry was an exciting period of change and discovery. The world then morphed, plate tectonics changed geology and our view of the earth, computer mapping, the replacement of secretaries with computers, computer graphics, the Internet, exchange of data and ideas, 3D seismic interpretation of oil prospects, and on and on, changed our ideas—and impacted the search for hydrocarbons. There is always something new to learn. The green energy search will change the energy industry of the future.

A partial list of products made from Petroleum (100 of 6000 items)

One 42-gallon barrel of oil creates 19.4 gallons of gasoline. The rest (over half) is used to make things like:

Solvents	Diesel fuel	Motor Oil	Bearing Grease
Ink	Floor Wax	Ballpoint Pens	Football Cleats
Upholstery	Sweaters	Boats	Insecticides
Bicycle Tires	Sports Car Bodies	Nail Polish	Fishing lures
Dresses	Tires	Golf Bags	Perfumes
Cassettes	Dishwasher parts	Tool Boxes	Shoe Polish
Motorcycle Helmet	Caulking	Petroleum Jelly	Transparent Tape
CD Player	Faucet Washers	Antiseptics	Clothesline
Curtains	Food Preservatives	Basketballs	Soap
Vitamin Capsules	Antihistamines	Purses	Shoes
Dashboards	Cortisone	Deodorant	Footballs
Putty	Dyes	Panty Hose	Refrigerant
Percolators	Life Jackets	Rubbing Alcohol	Linings
Skis	TV Cabinets	Shag Rugs	Electrician's Tape
Tool Racks	Car Battery Cases	Epoxy	Paint
Mops	Slacks	Insect Repellent	Oil Filters
Umbrellas	Yarn	Fertilizers	Hair Coloring
Roofing	Toilet Seats	Fishing Rods	Lipstick
Denture Adhesive	Linoleum	Ice Cube Trays	Synthetic Rubber
Speakers	Plastic Wood	Electric Blankets	Glycerin
Tennis Rackets	Rubber Cement	Fishing Boots	Dice
Nylon Rope	Candles	Trash Bags	House Paint
Water Pipes	Hand Lotion	Roller Skates	Surf Boards
Shampoo	Wheels	Paint Rollers	Shower Curtains
Guitar Strings	Luggage	Aspirin	Safety Glasses

Americans consume petroleum products at the rate of 3.5 gallons of oil and more than 250 cubic feet of natural gas per day! But, as shown here, petroleum is not just used for fuel. (Ranken-Energy.com)

My grandparent's home (during their retirement) in Greene, New York, located one block off Main Street.

Greene was first settled in 1801 when Indian lands were opened. This house was constructed in 1868 and many of the houses in town were built in the mid-1860s.

◆ CHAPTER 1 ◆   My Beginnings

Growing up in a small town in upstate New York

I came into this world just before World War II. My parents met in Brooklyn, New York. Mother was a nurse working at a Navy hospital and my father was just ending his first tour in the Navy. They moved upstate to my Mother's family farm and were soon married.

Grandfather Maurice English was a dairy farmer, but Grandmother Anna raised chickens. This pursuit of hers began so she could have a few fresh eggs, but soon developed to the point where her little operation required a large chicken house, which my grandfather built for her. She talked my dad into the chicken business as she needed help—grandfather had his cows and really disliked chickens. Soon, though, there were brooder houses filled with chickens scattered all over the meadow and a large additional chicken house was also constructed.

My grandmother, Anna Justin English, was raised in Albany, New York, and was the child of German parents who had migrated to the U.S. in the late 1880s. Anna had five sisters, but was the adventuresome one of the family. She worked as a governess and saved her pennies, and attended the World's Fair in St. Louis alone. She also visited cousins in Chicago, and finally traveled to Greene, New York, which is a small hamlet of 1,000 people located in the Chenango Valley in the central part of the state—and it was here where she met my grandfather.

They had three dates, one of which they spent alone together trapped in a carriage during a rainstorm. Three months later, my Grandfather took the train to Albany and asked for her hand in marriage.

They spent their first year of married life with grandmother's parents in Albany. Grandfather learned cabinetry from Anna's father and he worked in construction building new homes. But at heart, he was a dairy farmer, and he wanted to go back to Greene. Anna was a city girl, but happily followed my grandfather to the farm.

Their first farm was 80 acres and located next to Maurice's parents farm, and was situated about five miles west of Greene. He sold it after a couple of years and bought their second farm, located a bit closer to Greene, but to the east of town. It was a great spread with a marvelous house which included a glassed-in front porch running the width of the house. He soon determined though that it needed a new and larger barn to accommodate his growing herd of milk cows. But my grandmother put her foot down, "I will not move into a house with no running water and no bathroom!" So the building of the new barn had to wait for indoor plumbing in the kitchen and for the bathroom to be constructed including fittings for the sink, bathtub, toilet, and a septic system.

Their third farm was situated at the edge of Greene and it prospered in spite of the Depression. Milk prices were low, but expenditures were also low. They raised much of their own food. Grandmother sold a few excess eggs. They cut ice in the wintertime from the cattle pond, which had catfish living in it. The ice cooled the milk, so their electric bill was very low. The ice house was attached to the main house and led to the kitchen. Sawdust covered the ice and did not melt all year. In the heat of the summer, the ice house was a cool place to slip away into—you could escape the stifling summer heat, chip off a piece to eat, or add an icy chunk to your glass of water or lemonade. Ice was used in the kitchen refrigerator up until the 1940s when a modern refrigerator was finally purchased.

By the mid-1930s, more people were leaving Germany, especially Jewish families. Two such families arrived in Greene at that time and my Grandmother housed them for a while on their farm. She taught them English as she was bilingual. She helped one family buy a flat-bed truck and they began hauling live chickens to New York City, 200 miles away. They made a lot of money during WWII, much more than my father or grandmother who worked solely on the raising-chickens side of the business, as well as in the dairy business.

Yours truly, little Tommy, arrived in March 1936 and sister Jorette arrived in October 1937. By 1940, the chicken business was no longer making any money for my parents. At that time, my father switched gears and took a government job as a carpenter in Maryland, building barracks for a new Army base at Fort Dix. Roosevelt knew we were being pulled into the war and he started building bases. We moved to Maryland just outside the base.

One of my first memories was riding in the car on December 7, 1941 and hearing of the bombing of Pearl Harbor on the car radio. My sister, Jorette, and I—ages four and five—knew nothing of war or the implications of that fateful day's bombing—but we understood that our parents were exceedingly upset. I recall Jorette began crying and they tried to console her.

We moved back to Greene shortly thereafter and my father was called up as he was in the Navy Reserve. He was to be in a newly-formed construction unit, which was later known as the 'Seabees.' Mother stayed in Greene and sold off the chicken business, and returned to nursing. She refused to move to the west coast with my father, wanting to maintain a real home for her two children. During this protracted separation (as the war was long), my father fell in love with another woman, and my mother became a divorced woman. When asked by other people about my father, she always replied, "He never came home from the war!" People assumed he had been killed in the war. As a result of this implied cover story of hers, I don't ever recall anyone asking me about my divorced parents.

Mother worked in the small local hospital and soon became its manager, a position she held there until she retired. The shifts were 7:00 a.m. to 3:00 p.m. or 3:00 p.m. to 11:00 p.m., and sometimes she had to cover the 11:00 p.m. to 7:00 a.m. shift. She was often called for the birthing of babies and probably delivered more babies than some of the doctors. The consequence of these demanding hours of hers was that Jorette and I had to prepare our own breakfasts and make our own lunches for school. Sometimes we cooked dinner too as Mother was tired from her shift or possibly working a double at the hospital.

The rules at our home were simple. We never left the house with an unmade bed. No clothes or towels were left on the floor. There were no unwashed dishes (we had no dishwasher in those days of course). The furnace was set on low to preserve heating fuel. No lights were left on. The house was locked (the key under the rug on the porch—I wonder how many robbers never look under the rug).

Ours was the first house outside of the town boundary so we could take the school bus or walk with our friends to school. It was more than one-half mile to the schoolhouse, but we usually walked. In later years, we rode our bicycles. Life was busy, life was good, and there were always odd jobs to be had for the asking so we could earn a bit of pocket money.

In 1947, Kay McNulty and her family moved to Greene, New York when we were in the fifth grade. She and I both got our growth spurt early on and were the tallest girl and boy in our class all the way until our early high school years. Possibly that's why we first became friends. Kay was a good-looking girl with long dark straight hair, and brown eyes to match—she was smart as well as ambitious.

School cultures were stratified into several different groups. There were 55 of us in our class, split into two classroom groups, which shuffled around from year to year. The top four students were three girls and one boy, and they studied more than the rest of us and were also very competitive about it. From there on down to the bottom of the class, there was less competition. Cousin Dick and I were number five and six grade-point wise, Kay was seventh or eighth in the class.

Then we had the sports jocks, who were lucky to play as they continually slipped into probation. The girls were big into cheerleading and twirling batons. Then there was the school band which was very active and had students from several grades. Kay chose to play in the band—a string base. I tried sports: football for two years, track for three years, and wrestling for four years. I was never good at any of them.

Kay's father, James (Jimmy), was the second in command in the Sales Department for Ansco Film Company in Binghamton. Jimmy got Kay and I started doing our own printing. I set up a darkroom in our basement and we spent many hours together there (although, admittedly, not always printing). We tried out every new film and paper that Ansco put on the market. We even did color work before most people really got into color photos. I built my own enlarger with a lens donated by my science teacher. Finally, I saved enough money to purchase an enlarger.

Jimmy got things done and everyone spoke to him rather than to his boss. They sold their home in Binghamton in 1947 and bought some property just north of Greene.

It was the beginning of summer and they had two months to get out of their house and complete their relocation. It was a 20-mile commute to work for Jimmy who took a couple of weeks off from work, hired a hand or two, and built a large 20' x 40' open building. A bathroom was screened into one corner, the kitchen appliances in another and the rest was outfitted with beds and chairs. Elsie, Kay's mom, was a teacher as well as the librarian for a Binghamton school. The McNulty family had five children—Mary, Kay, Jerry, John, and Gene, and a dog plus a cat also, as I recall. The family lived in this makeshift dwelling for a couple of years until Jimmy built a house on the property.

I occasionally spent a few hours helping build the family's final structure on weekends—mostly on Sundays after church. Kay or her father would invite me over for barbequed chicken (really as an incentive to work on the house). Kay, Jerry, and I would pound nails or whatever in pitching in to help Jimmy complete the family's house. It was a long, rambling one-story flat-roofed building with five bedrooms. The outside was stucco and we kids painted it white. The family moved into it long before it was finished. The sheetrock wasn't in place, the floors were cement slabs without tile, light fixtures consisted of single bare bulbs, wires hung down—the septic didn't work right, and Jerry and I spent what-seemed-like-forever digging a hole for a new septic tank…by hand.

Jimmy was a beer-aholic as every evening after he got home he drank a 12-pack of Utica Club beer. Kay, Jerry, and I had many a beer with ole' Jimmy. No one in the McNulty family told us we were too young to drink. I think my mother and grandmother must have been charter members of the WCTU (Women's Christian Temperance Union). My mother might catch a whiff of beer on my breath when I returned home, so I became very careful in avoiding her. She told me I could drink beer at home when I reached the legal drinking age. On my eighteenth birthday, I placed a six pack of beer in our refrigerator. I was in the next room when she opened the refrigerator—I heard a large gasp, but she never said anything about it.

Their barbeque was a long rectangle of cement blocks—two or three tiers high—with a metal screen on top. Jimmy's barbeque sauce was good and they always marinated their chicken overnight. We chowed down on his tasty barbequed chicken, grilled corn, sometimes a salad, and washed it all down with lots of ice cold beer!

Pursuing an education

Kay and I dated through high school—and lost our virginity with each other. In 1954, I went off to the New York State Merchant Marine Academy in New York City. She went to Cortland State Teachers College and finished in three and a half years. I dropped out of the Merchant Marine Academy after the first year—the military routine wasn't my cup of tea. The US Government gave up supporting the Merchant Marine at this time and it looked to me like opportunities for jobs were disintegrating as a result.

Then I had to make a decision about a different college to attend. Kay was probably an influence and wanted me to study somewhere locally. Three days before fall registration, I went down to Endicott (part of the tri-cities area of Binghamton, Endicott and Johnson City) and visited Harpur College, which was a newly-established resource by New York State for veterans who'd served in WWII and Korea. At that time, the school had less than 1,000 students. They accepted me because I had a New York State Regents Scholarship. I had no major initially but knew it would be in the sciences. I picked geology as a major when I was a junior as I had already had taken two courses. It was a natural fit for me as I had already studied geology on my own and even mapped some of the local geologic features around Greene. Kames, kettles, moraines, glacial erratics, and eskers were part of my vocabulary long before I took a course in Glacial Geology.

Continental Glaciers move over the land downhill simply by the weight and thickness of the ice. It is thought they were over one mile in thickness when they were moving, and at their maximum mass. The movement scraped off the surface of the land and plastered glacial moraine over the low places. In some areas, this ice gouged out the valleys, cutting them as much as 1000' deeper—which led to the formation of the Finger Lakes in western New York State. When the glaciers melted, they left glacial moraines and glacial till on the surface. Streams flowing off and out of the glaciers left gravel deposits (kames), and perched stream deposits (eskers). Glacial erratics are large pieces of bedrock left scattered over the land surface. Kettles are lakes formed by melting ice blocks.

Tying the knot

Kay and I didn't see each other very often during this next period of time when I was at Harpur—she was 40 miles away at Cortland without a car and I was commuting 30 miles to school plus was working at the local grocery store in Greene (25 hours per week, sometimes up to 40). She got a ride back to Greene on weekends sometimes—and I, in turn, made some trips to events at Cortland State.

Everyone, especially Kay's mother, Elsie, expected us to marry. I don't remember actually proposing to Kay—although her mother kept pushing for us to tie the knot during Thanksgiving vacation, while I wanted to wait until June when I finished college. I remember going with her and purchasing an engagement ring with a small diamond, probably just a short time before we were married. (Kay claims a year before—maybe it was!) But you can tell who won that debate as we got married on November 30th, 1957, with Kay graduating early the next year, in January.

My grades were terrible, especially considering I wanted to go to graduate school. I was working too much to afford college, and was spending too little time studying for classes which I just wasn't interested in. I lost some credits in transferring colleges, so I then went to night school for two calculus courses and attended two different schools that summer—taking Spanish at Alfred University, and music appreciation at Colgate University. Not surprisingly, graduate schools weren't interested in taking me given how I looked on paper. However, the Harpur President, Dr. Glenn Bartle, was a retired oil geologist who regularly visited our Geology department and he heard two or three lectures I gave at senior seminars. It was following one of these sessions that I told Dr. Bartle of my problems. He said, "I'll recommend you to my school and old friends at Indiana University." And I subsequently overheard one of his promised phone calls, saying to the person on the other end of the line, "I want you to take my boy, Tom Cochrane. He works hard and is a good student."

And that's just what happened—so we packed up and moved to Indiana with graduate school finally at hand.

Kay was hot to have children—I wanted to wait until I finished my PhD. In spite of being good Catholics, we used limited contraceptive methods which proved to be ineffective. As a result, our daughter, Maureen, was born in February during my second semester at Indiana University. I got a low-paying assistantship position during the second semester which helped a little. By June though we were broke and I had little chance of a real paying job for the summer. We returned to Greene looking for some seasonal employment opportunities. While there, we stayed at my mother's house as my sister Jorette was away at college by then.

Becoming a teacher

I was trying for any type of scholarship or assistantship back at Indiana University but nothing was developing. Kay's mother had plans for us however. "Why don't you teach school here in Greene?" she inquired. I wasn't even going to consider it. The grade school principal lived next door to the McNulty's, the high school principal was a member of the Catholic Church, plus their family socialized with him, and he was also my mentor and teacher in studying science.

So Elsie took it upon herself to set up interviews for both of us and I agreed to go and talk with the high school principal, Andy Pearl. We got there for our interviews and Bob Bennett, the grade school principal, took Kay off to look at a sixth-grade classroom. Mr. Pearl then escorted me off to the Science Department, telling me the chair of the department had retired, here was my lab and informing me I would be teaching 7th & 8th grade math and 7th & 8th grade science, plus 9th grade science. I would also be named acting chair of the department. He then told me that in 15 minutes Kay and I would be meeting with the Superintendent of Schools to sign our contracts. (Wow! Talk about railroading a person(s) into a job! We had been had by big Elsie. She was a prolific volunteer for everything there locally — which meant she ran things while arranging for other people to actually do the work.)

Kay loved teaching, but hated working and raising children at the same time. She loved kids — and wanted five, just like her parents.

Four years passed. Kay worked on and off during the birth of two more children. I had no training in Education in terms of actually learning how to teach, so I attended summer school at Colgate University for three summers. Our salaries were low and money was tight.

I wanted to save money to go back to graduate school in geology. In my current career, I continued to teach 9th grade Science, Earth Science, and Physics. I was now the Chairman of the Science Department (there were three of us), and acting Chairman of the Math Department as the Chair was ill. I also taught Driver's Ed after school and in the summer to earn a few more bucks. I even became an assistant wrestling coach as I had wrestled in high school.

My fourth year teaching, I also worked 28–30 hours per week for my cousin, Dick Capra, on his farm. My schedule was to get up at 4:30 a.m., drive to the farm and help milk the cows, drive home, take a shower, get dressed, and get to school by 8:15 a.m. After school, it was back to the farm, spread the manure from the day, and then help milk the cows. Dick and I drank a lot of wine during the milking. Home by 8:30 p.m., shower, and bed. I did that for a year before I came down with a serious sinus infection. That was the end of my farming experiences, but you know what they say, "A boy from the farm is always a farmer at heart."

About this time, I received a National Science Foundation Grant for high school teachers to study Glaciology on the Juneau Icefield in Alaska—I was in seventh heaven! One of my geology friends, Chris Egan from Harpur, had also received an NSF Grant. We picked up a third new friend, Mark, and decided to drive my VW Microbus to the land of the never-setting sun, at least in the summertime.

We left three weeks early to give ourselves enough time for the long journey, and I told Kay to meet me in Indiana with the children in September as I was planning on going back to graduate school. We three geologists had a marvelous time. We took two days to climb Middle Teton, sleeping part way up on the mountain. To camp at over 11,000' elevation with the stars appearing so close you could almost reach up and grab them was a wonderful experience. We left our names at the top in a designated box to mark our accomplishment.

In visiting Yellowstone, we encountered bears in the road and moose in the meadows. We cooked steaks at a campground there and a large bear wandered in, eager to snatch them right off our fire. We yelled at him, but to no avail. I picked up a large branch about 15' long and beat it on the ground in front of him yelling, "You can't get my steak!" He then left, but that was a pretty stupid thing I did.

Next, we climbed a mountain in Montana. I was on a sheer cliff with a rope, but the wall of rock was a conglomerate. The round rocks kept coming loose from the rock face, so it took me three hours to get off that face. At this point, I decided I would give up serious mountain climbing.

Finally, we drove up the Alaska (ALCAN) Highway. It was a gorgeous but dusty ride. I had two spare tires and turns out we needed them. Somehow, out in that vast wilderness, I picked up three nails in my tires at different times on that trip. Luckily we had the two spares and were able to find a gas station to repair them before I hit the third one.

We stopped at a hot spring next to the road and had a swim. On the way back, we also stopped and were surprised to see there was snow around the hot spring, but the water was still piping hot, with steam rising up into the nippy air. We drove to Haines Junction and then ferried down to Juneau. The three of us were flown, as an advance team, by helicopter into the ice field camp on the Taku Glacier (I'm not sure how we three were selected for the advance team—maybe it was just because of our enthusiasm!). That was a bit harrowing though as the weather socked us in and it was nine days later before the rest of the 25-person group, including instructors, landed in a National Guard plane equipped with skis.

We spent those nine days working on getting our equipment working. We changed the oil on snowmobiles, gassed 'em up and started them—however we could not get the main generator started, which would have provided us with radio contact with the group in Juneau. We took one of the snowmobiles up the glacier from Camp 10 (which was the main facility) to Camp 9. We picked up another snowmobile there and

Supplies were delivered by helicopter and the Alaska Air Guard (1965)

took it back down the glacier to Camp 10. We were glad indeed when the main group arrived, bringing with them fresh food, as all this while we'd survived on old WWII K-rations.

This was a marvelous experience. I got to see how glaciers deposited the glacial features that I had grown up with around Greene, New York. We lowered ourselves down crevasses and counted the layers of ice from many years back. We collected samples for NASA which were slightly radioactive as they contained fallout from Russian nuclear bomb tests done in Siberia in 1959. I climbed a substantial mountain (11,000' elevation) which had never been climbed before to our knowledge. The National Geographic sent William Garrett to photograph us for an article on glaciers. He spent at least one day with my group of three as we shot a survey line and set out stakes to observe how

That's me in the front (in my Greene sweatshirt) listening to our leader, Dr. Maynard Malcom Miller. The water bottles on the shelf were samples we collected for NASA which had low-level radioactivity—but we didn't seem to be very worried. Photos were taken by William Garrett, National Geographic Photo Editor. Source: National Geographic Magazine, June issue, 1965 different parts of the glacier moved. We ran a gravity survey to show the shape of the valley beneath the Taku glacier. As a result, I appear in the group photo of us in a 1965 National Geographic issue, wearing my (green!) Greene High School sweatshirt.

At the beginning of September, we drove the 4500 miles back in four and a half days. Not surprisingly, my poor VW bus needed a valve job after the trip!

Back to grad school

Kay arrived in Indiana with a U-haul trailer attached to one of her brother's cars. I should have felt guilty, but she was the one who wanted all the children and prevented me from getting my education. The next academic year went well. I had an assistantship which mostly involved dark-room work and taking pictures of various minerals under the microscope for a professor who was doing a study on low grade metamorphism in a region of New Zealand. He had hundreds of rock thin sections and wanted me to pick specific slides and find certain minerals for him. I took hundreds of photographs in searching for what he was seeking and succeeded in providing the proper view he required, and his article was published in the G.S.A. (Geological Society of America) bulletin.

At this point, I applied for assistantships in other colleges and got an offer for a position at Ohio State which had a good department encompassing glacial geology. But Kay had other ideas, and once again went off the pill and got pregnant. Gawd! I don't even think we had sex that often as I was always studying. I was angered by this pregnancy, having once again been excluded from the decision-making, and our relationship never recovered.

[Note: "The pill" was not in widespread use until 1962 and Maureen and Chris were conceived before it was on the market. We used other methods of birth control—which obviously were not very effective. Although Elisabeth and Brian were conceived when I thought Kay *was* on the pill.]

Here's a helpful Oilfield Glossary for additonal reference courtesy of Schlumberger:

www.glossary.oilfield.slb.com

◆ CHAPTER 2 ◆   My life in the oil business begins

How did a glacial geologist become an oilman?

So…where to find a job using my geologic background? I could easily get a new job teaching high school math and science—I just needed one more summer at Colgate to complete a Master's degree in Education. I had enjoyed the four years of teaching at Greene Central School, however there was simply no money in teaching. (I began teaching at the magnificent salary of $4,000 per year, and after four years my compensation was still not up to $5,000 per year, which is why I needed to supplement my low wage by working summers as well as part-time after school.)

Ah, what to do? Why an oil company?

In actuality, the greatest numbers of geologists work in the oil and gas industry. Additionally, a few geologists work in the mining industry, a few work for NASA studying the moon, Mars, and beyond, some work in the field of oceanography, some for State Geologic Surveys, and a few work at local levels for municipal planning departments.

So, back to my plight in 1964—the need for a good paying job. What does one put on his résumé to qualify for a position with an oil company? BA degree in Geology from Harpur College (with no course in petroleum geology), two years of graduate studies at Indiana University (with no course in petroleum geology), a couple of assistantships filing maps and fossils, one assistantship taking micro-photographs of thin sections of metamorphic minerals (of no use in the oil industry), previously a stock boy in a grocery store and, last but not least, knowing how to milk a cow! Oh yes, four years of teaching junior and senior high school science and math, plus teaching Driver's Ed, being an assistant wrestling coach, and three summers at Colgate University studying Education courses. Wow! I should have no trouble getting a terrific position with them given that impressive body of work!

At the time, the oil companies were visiting universities and interviewing geology and geophysics students—even though it was a slow time in the oil industry. It seems the interviewers did not want to be seen in the office as their companies were actually laying off staff! My situation was drastic, my resumé had been submitted everywhere to companies and governmental offices. I suffered through what was finally 35 interviews in order to get a job.

Most of the interviewers were middle managers, mostly nice guys about to be retired early by their companies. I quizzed them extensively about their firms and got a real sense of some which I would not want to work for. But I was also getting no job offers, so that was a moot point. By interview 34, I was so disgusted I didn't care what I said to whomever.

Interview 34 was with the Division Geophysicist, a man named Ken Gilbert, who worked for Pan American Petroleum (later changed to AMOCO, now part of BP). Reasonably enough, Gilbert wanted to know my qualifications for the opening. After explaining my admittedly meager background, he indicated he wanted me to have a PhD in Physics and one in Mathematics. I told him he was barking up the wrong tree—that no one in the Geology Department had that kind of a background. I left the interview really pissed.

Previously I had scheduled interview 35 with the Division Geologist for the Oklahoma City office of Pan American Petroleum. After his interview, Gilbert took me to meet John Mason—who looked like a real clown wearing cowboy boots and a sports jacket which looked like it was made of awning material. I laughed when I saw him with his feet on the desk, and demanded he tell me why they wanted a geologist instead of a geophysicist, since most oil is now found using seismic tools instead of basic geology. I hit him with my conclusion that geology in the oil business was a dead end and outdated by geophysics! (Remember, I had no experience or knowledge whatsoever of what went on in oil companies.) After our lengthy interview concluded, John had missed the next four applicants he was scheduled to meet. Two weeks later, I received an offer of a job with Pan American in Oklahoma City. I began working for Pan American Petroleum as a Jr. Geologist with a starting pay of $7,200/year, a significant step up from my teacher's salary.

Launching into my new job—and career

So off I went to Oklahoma City, with my wife and kids in tow. The new Junior Geologist with no background in oil and gas was suddenly dumped into an office with 15 other petroleum geologists, each of whom had received their geologic education from oil schools in Texas and Oklahoma.

The new kid at the company, day one

My letter of employment indicated I was to report to Howard Cotton, the District Geologist for the Oklahoma City Division of Pan American Petroleum, so I duly reported for work at 8:00 a.m. sharp and was directed to Mr. Cotton's office. His secretary says, "Howard's in the usual Monday morning district meeting and will be in his office at 9:00 a.m. Have some coffee and wait in his office." About 9:15, Howard shows up and is told that I'm waiting in his office. "Oh! So, *you* are the new geologist. I thought you were coming in September." I reply, "No, my letter says to report to you on June 15th."

Howard, a short, middle-aged, and slightly paunchy man with a slight drawl then says, "Well, we'll have to figure out something to do with you. There aren't any free offices on this floor. Maybe we can put you down on the 14th floor. I'm leaving Wednesday for my month-long vacation, so I can't really deal with this until I get back. I suppose you went to Oklahoma or Texas for your studies?" I reply, "No I got my undergraduate BA from Harpur College in upstate New York."

"What is Harpur College?"

"It's the liberal arts branch of the State University of New York (SUNY)."

"Egad, you have a BA? Do they teach geology there?"

"Yes, they have a very good geology department. I also have two years of graduate work in geology and geophysics at Indiana University."

" Did you study petroleum geology there?"

"No, I haven't had a course in petroleum geology."

"What were you majoring in?"

I reply, "Glacial geology and glaciology."

Howard then asks, "What is glaciology?"

"It's the study of current day glaciers. I spent last summer on the Juneau Icefield in SE Alaska. Glacial geology studies past glacial periods and deposits."

"We don't have any glaciers or glacial deposits in Oklahoma! Who hired you?"

"John Mason."

"Gawd, I might've known! He's always trying to put something over on me! What's next? A glacial geologist!"

I then told him, "I have had courses in stratigraphy and sedimentation, sedimentary petrology, and geophysics which apply to petroleum geology. I had practical field usage of geophysical instruments including magnetometers, gravity meters, and surface resistivity. That should be of some help. I am also a willing learner."

Howard then retorted, "Well, we'll find something to do with you, but I feel you will not last long here in this office. For now, take Ron Eddington's desk as he's on three weeks' vacation. When I get back, we'll work something out. Report to Mike Maravich and he'll get you out to a drilling well so you can actually see what we're about."

I leave his office and the secretary snags me, saying, "Did you get all the company paperwork done? Let me take you down to the Personnel Department. Howard is more bark than bite. You'll be alright." She then took me around and introduced me to a few of the other geologists.

I think to myself that I will never fit in here, especially with this asshole as a boss. I wish I had been on the icefield this summer—especially now that I find out no one wants me here for months yet.

They still went through the process of looking around for a desk for me—but indeed there were no empty offices, just the one guy on vacation—so I go ahead and take his office. Two weeks later, Ron Eddington comes back to work and finds he now has no office. So instead of giving me a desk in the hall, they give me Ron's office and put him in the hall—I wonder to myself who he had pissed off! Ron took it well though, and he and I even became good friends.

So what does one do with a new geologist with no experience? Well, first, the field coordinator, Mike Maravich, takes me out to see a company drilling well. I had never seen a drilling oil well before. I had seen water wells drilled with small truck-mounted drilling rigs. These drilling rigs were big, and noisy, and smelled of diesel. They pulled 90' of pipe out of the hole at a time, so the top of the rig jutted 120' up into the air, or taller yet. Mike give me a new binocular microscope and a fluoroscope (I'd never seen one of these; oil samples fluoresce when placed under ultraviolet light), and told me to look at drilling samples to see the pay zone that the well was about to cut through. The drill bit grinds the samples up into small pieces about 1/16" in diameter. Over 90% of the sample caught by the drilling crew is shale sloughed down the wellbore from uphole. I had never run samples before (I thought geologists just looked at hand specimens of real hunks of rock—not ground up dirt! Wow, I guess I have a lot to learn!)

Obviously, I was going to be a problem. So they then turned me over to Erwin "Doc" Selk, who was the sample expert with the company, and one of the few people who lasted to retirement with the company, at age 65. Doc Selk barely had a couple of years in college and had instead learned his trade working on drilling rigs. He made extensive and beautiful sample logs and taught me how to find oil and gas pay zones from samples alone.

A couple of years later, the company sent Doc Selk and me to the Illinois Geological Survey to assess the possibility of developing a new play in an old oil area there. In five days' time, we ran over 50,000' of samples from bags that were saved in five or 10' sections per bag. The goal was to examine the top section of the St. Louis Formation for porosity and oil stain possibly left in the samples. That formation had not been the target of the wells drilled, but could lead to an oil play in a zone previously considered as too low permeability to produce oil. Fracking could enhance the permeability and make the formation a target for overlooked oil production. To coordinate our work into a similar work product, we both ran samples from several of the same wells for comparison. Doc picked a formation top in one well that I could not find. I made him show me how he picked the top. It turned out there was only one small chip of the formation in the sample bag which he picked the top on. (I thought, Wow, I'll never get as good at this as Doc!)

My next lesson followed when I was sent out to a well to pick up the electric logs that were run when the well reached total depth. Of course, I had never seen electric logging performed on a well and had never had a course in electric log interpretive readings.

I quizzed the Schlumberger Logging engineer for his analysis of the potential pay zones. He gave me several books on log analysis and signed me up for a logging school at their offices. Electrical surveys measure several different rock and fluid characteristics. Gamma rays measure the radioactivity of the zone—shales have higher radioactivity than sandstones or limestones. Neutrons bombard the rock for porosity and to determine the types of fluids in the rock. Resistivity is measured: saltwater has low resistivity, oil has high, etc. The drilling fluid (mud) is weighted to hold back the formation pressures—we don't want the wells to blow out. Over pressure puts fluid from the mud into the rock formation and leaves a layer of mud (mudcake) coating the wellbore. This gives us an idea of permeability. There are logs to measure the dip of the rock formations, logs to indicate natural fractures, etc. It's a complex study with lots of interpretation based upon the geologist's knowledge of the kind of rock formation, salinity of the formation waters, and many other factors.

Doc Selk Sample Log

Note the attention to detail.

The sample log looks at Porosity, Lithology, Size of Grains, Cement, and Accessory minerals. The Remarks section describes the sample, and identifies formation tops along with oil shows if noted.

It loses a bit of impact in this black/white printing but still illustrates well the kind of detail someone like Doc Selk could see in samples.

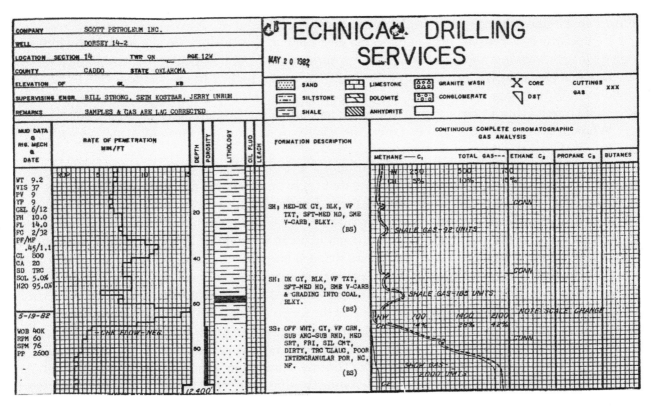

Gas Unit Mudlog from Technical Drilling Services, prepared by the wellsite geologist. Note the small gas spike from a black shale and the gas spike from a possible gas-bearing sand.

Well, I picked up the electric logs and headed back to the office, two hours away. I got back and am asked for my recommendation—should we complete the well or plug it as non-productive? I have no clue as I only have the Schlumberger engineer's analysis. It probably is productive. So, next, they send me up to have the Division Logging Engineer's interpretation. The secretary says, "Watch out for Doc Holland, as he's often high on pain pills."

I take the logs into his office late in the afternoon and introduce myself as the new junior geologist. He starts asking me questions, "Is it a company well? Is it a development well, or a wildcat? How much acreage do we have around the well? Have you picked the formation tops (elevations)? How does it fit the maps for the area?"

Of course I cannot answer any of these questions—and immediately go on the offense retorting, "What difference does it make? Don't the electric logs give the same information—no matter what the deal is or the formation measured?"

The conversation breaks down from there and Doc Holland sends me off to find out about the well, the deal, the maps, etc. I worked late that night and the next morning I'm back in his office, not possessing all the answers, but some of them. I had marked the electric log formation tops and plugged the well data into structure maps made for the well recommendation. I had read the well recommendation report and made my own conclusion as to whether the well would be productive or not. Holland was impressed with my desire to learn and we became good friends thereafter.

He tended to gain weight easily, so every four months or so, he would take "uppers" to lose weight. These were his productive times. His job entailed talking to everybody in the company as well as in the industry all over the world. He had the largest phone bill in the company. (He invented the grapefruit spoon and sold it to Oneida flatware and gave me one which I still use and treasure—Wikipedia does not mention his name though, so maybe he was putting me on.)

Next, I was assigned to work with Dr. Tommy Thompson, who had convinced the company to buy 120,000 acres of leases covering parts of four counties in central Oklahoma. The target was the Mississippi Solid, a dark dense rock which produced oil and gas from natural fractures.

Pan Am cut cores over 500' in length in 10 wells in Kingfisher County. I got to see the coring and examine the rocks with Tommy Thompson. The project also involved computer mapping in the Research Department in Tulsa, Oklahoma. It was a new project and we were involved in the first computer maps made by the company. We wore white lab coats and booties over our shoes in a huge cold room filled with computers with spinning discs. Tommy and I picked all the electric log tops on about 2000 wells which were fed into the computer program. We made all kinds of maps. We had a peek into the future of computer mapping. It was a terrific learning experience!

The wells turned out to be less economic than projected. (Imagine the costs of the coring and the Research lab computer maps!) The company then decided the play should be farmed out for as many wells as they could get drilled. Dr. Tommy Thompson was assigned a new project while Tom, the junior geologist, was directed to oversee the farming out aspect, with Pan Am retaining a small overriding interest in the wells. So yours truly was left with that loser area and subsequently left alone as everybody else then abandoned the project. (How interesting in retrospect that once higher management condemned a project, no one will even talk about it, or for that matter want to admit they had a part in recommending it. Think of the millions of dollars Pan Am spent during three years of putting the play together plus another two years of my involvement in farming out the acquired acreage. The industry, particularly the smaller companies in subsequent years, drilled thousands of wells for fracture reservoirs—and made money at it!)

Geologists pick formation tops off electric logs. Different formations have distinctive wiggles on electric logs—a shale formation has a high gamma ray signature, a limestone is very resistive, a porous sandstone has a large SP (one of the curves) and has low resistivity from saltwater (unless it has oil or gas in it). We make structure maps on various formations and chase oil and gas updip out of water-bearing zones.

Nearly every day I took 10 deals to farm out to District Committee. There were about a dozen horizons which produced oil and gas in offset wells and many small independent companies were eager to drill the wells. I read every report in the file and talked with all the geologists in the company who had written reports on any part of the area. In talking with them I was then chastised for never being in my office. In two years' time, I met everyone in the industry operating in that region of Oklahoma. Every deal I thought was a good economic project I recommended in District Committee. The answer was "Farm it out. It's not what we're looking for." The farmouts made several very successful oil companies and I learned the oil business. Later, these contacts would prove very valuable.

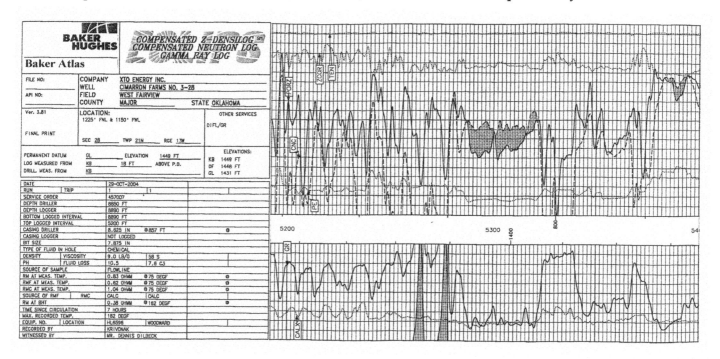

Baker Hughes Electronic Log

Density/Neutron Log shows porosity indicated by the shaded section of the log. When the neutron curve is at its lowest point of the density curve (with the readout itself displayed sideways as above), gas is indicated. The curve on the bottom is a gamma ray. Note the two spikes in thin shale zones. When the gamma ray moves up, it indicates a shaly zone. Clean sands or limestones move the curve downward.

When I think back on the writing skills needed for authoring scientific reports, mine were essentially zero. I had not done a thesis or any papers at all in any of my science courses. Instead, I had only written one long report on my geologic studies at Geological Field Camp in Wyoming. During my four years of teaching, I wrote nothing. What I was good at was taking notes. I was a speed reader, but a critical speed reader in that I could quickly weed out the pertinent data from a report. All the other geologists hired by Pan Am in those days had PhDs and had written a long scientific thesis in obtaining their doctorate. They all had mastered methods of writing scientific papers. Reports for the company were highly abbreviated with no abstract or introduction. They began with a recommendation and then the body of the report outlining the history of other dry holes or producing fields in the area, and highlighting why a new well should be drilled at a particular location. Only extensive regional reports were written with bibliographies or references quoted as, most typically, they simply wanted a brief recommendation for a well and the reasons why it should be drilled, period.

The PhDs started their research in an area by reading Scientific Journal articles and regional studies previously written by someone in the company. Once this was mastered, they would look at the records of individual wells.

My approach was to look at the wells first, especially the ones the company had already drilled. The questions were simple. Why was the well drilled? What was the target formation? Why was the well a dry hole? Where should the well have been drilled to find production? Most of this can be dug out of these earlier reports, but the quicker way is to talk with the originating geologist and the wellsite geologist who drilled the well. Their answers are not usually in reports. The company also never required a follow-up report on the results of the well—be it a producer or a dry hole.

I did have a good start on prospect evaluation in that I was assigned to work with Dr. Tommy Thompson who was a good scientific thinker, and a good teacher. I especially learned about coring wells and fracture type reservoirs from him.

What's "wildcat" mean?

The process of drilling for oil or natural gas in an unproven area, that has no concrete historic production records and has been unexplored as a site for potential oil and gas output. ... Wildcat drilling is also known as exploratory drilling.

—Investopedia.com

My first wildcat well and field discovery

Since I was identified with a loser project, I decided to work on a project outside of the area I was assigned to. I picked an adjoining area with a different target sand called the TWG Area, e.g. the area between Taloga, Watonga, and Geary, Oklahoma. Another division of Amoco was the Liberal District (previously stationed in Liberal, Kansas, and recently moved to Oklahoma City but it had not yet been merged into our district.)

The District Geologist was the afore-mentioned Howard Cotton, who was in charge of our office and who I thought didn't like me. Howard was a short, middle-aged fellow from the south, a 'good ole' boy,' a 'yes man,' and highly tuned-in to what his superiors wanted or instructions they had given—he could adroitly switch sides in mid-sentence if he perceived his boss didn't like a particular project. One day, he was making the rounds with the District Supervisor, Jake Warden, who did like me, so I showed them the project I was working on.

Howard said, "You're not supposed to be working in that area."

To which I replied, "Well, no one else is working on it. The Liberal District is drilling like crazy just outside of the area."

The District Supervisor then said, "Keep working on your project, Tom, and I'll talk to you about it later. We need to drill in that area." I'm sure this did not set well with Howard—but his boss was for the project, so Howard was pretty much stuck waiting to see what I came up with.

My project was an innovative one for a stratigraphic sand. I wanted to drill three miles east of the mapped Morrow Formation sand limit in a township with only one other well, a dry hole. I based this on my interpretation of a seismic low having more sand than the dry hole to the west which was located on a seismic structural high. I got the geophysicists in the District to substantiate my interpretation. There is much rivalry in companies between geophysicists and geologists. I had several disagreements in District Committee over geophysical interpretations which were not supported by the geology. This one they supported!

Finally, I took the project to the District Committee who approved it—with Howard again voting no. Normally the District Geologist took all the approved district deals to the Division Committee for final approval. Howard said, "Since this is the first well deal you've worked up (and a real wildcat), I want you to take it to the Division Committee for the experience of presenting your deal."

I believe he thought they would not approve it. But my mentor, John Mason, was the Division Geologist and occasionally we had drinks after work. John loved the project—he liked anything which was innovative and outside the norm. The project was approved and a wildcat allocated to be drilled at my recommended location. (Management often moves the locations to the center of the yellow—e.g. the acreage block the company owns.)

The Engineering Department decided to use a different mud system to drill the well, which made it more difficult to evaluate the samples. The well turned out to be the most expensive well the company had drilled that year in Oklahoma. Howard knew this would end development, even if the well was successful. The well encountered a 20' gas sand with nearly 10,000 psi pressure. The gas and oil production from the well paid for itself in 66 days at a gas price of 22¢ per mcf (thousand cubic feet of gas). Subsequently over 150 wells were drilled in the township, many of them by Pan Am.

If only I had been an independent consulting geologist with a 1% override fee!

At the time, no one seemed to make a big deal out of the discovery. Pan Am was drilling exploratory wells across three counties, finding gas fields in regular succession. I was not aware of the statistics for the success of wildcats versus development wells. This one was just one more discovery (e.g. what we were paid to do), and it was only after maybe ten more offset wells were drilled that the size of the discovery became apparent. Pan Am (e.g. Howard) switched me to a new project, so I only saw the detailed mapping of the field that was produced by other geologists and petroleum engineers in the company. I would have liked to have been involved in the mapping of the field as it developed.

In four years, I found over 300 billion cubic feet of gas for the company and they doubled my salary. I was bitten hard by the wildcat bug!

Why wells are fracked

Rock formations generally have low porosities and permeabilities. There are no pools below the ground, so the oil and gas must be coaxed out from the rocks. Initially, oil and gas are under high pressures from the weight of the overlying rock formations and the oil or gas will flow to the surface. Once the initial pressure is gone, oil wells are put on pumps, and gas wells may be put on compressors. To extend the producing area around the wellbore, many or most wells are given a frack job of varying size. The common fracture fluid is water, usually fresh water. It is pumped into the formation at high pressure to overcome the formation pressure. The frack water follows the path of least resistance, which may become a problem in shallow wells. The frack opens up naturally-occurring fractures and also creates more fractures yet in the rock formation. These fractures we want to keep open, so sand or glass beads are commonly added to the frack water to prop open the induced fractures. Chemicals may be added to the frack treatment process and are the source of much of the objections to fracking. Clays in the rock formations swell when fresh water hits them, so these added chemicals are used to prevent the swelling of the clays which would plug up the formations.

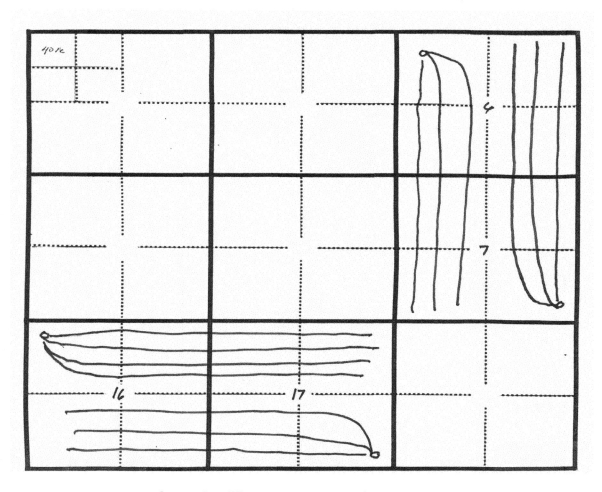

Horizontal Drilling Pattern to Drain Two Sections

Two inital wells were drilled from two well pads. Multiple horizontal legs were drilled from the two initial vertical wells. Multiple frack stages are performed along each horizontal leg. Fracks may use 3–5 million gallons of water, plus tons of sand to prop open the fractures, as well as other chemicals.

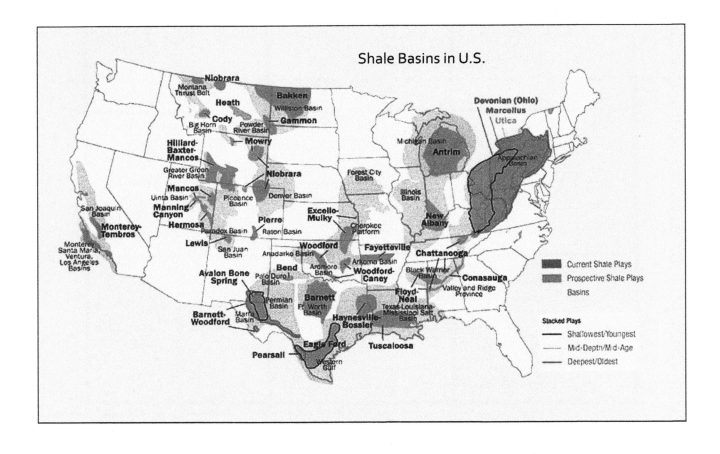

The current fracture plays in the U.S. are mainly in shales and very dense rocks like the Mississippi Solid. The practice is to drill horizontal wells in the target formation and give it a very large fracture treatment of thousands of barrels of treated water. The spacing in the deeper formations in Oklahoma is 1280 acres, which allows the horizontal section of the well up to two miles in length. The frack water flows back or is pumped back out of the wellhead and then the oil and/or gas flows out of the well. The industry is not currently reprocessing the frack water (which probably should be done). The frack water is then disposed of into shallow porous formations. This pressures up those formations, may get into the fresh water table, and is causing earthquakes.

Common waterfloods have for many years put produced waters back into the same formation, with little or no problems. The oil and gas is taken out of the highest wells in the formation and the water is put into the lowest wells. This does not cause over pressure in the formation, but replaces the produced oil with water.

The shale formations are such low porosities and permeabilities that the oil, gas and water have not separated into different legs. Hence the frack water is not put back into the same formation.

◆ CHAPTER 3 ◆ Where Oil and Gas are Found

So where do you look for oil and natural gas?

Oil, natural gas, and coal are organic deposits and represent previous life forms from other eras. For example, ferns and leaves are commonly found in coal deposits. Particularly during the Carboniferous Period, 300 to 369 million years ago, large marine swamps grew and were covered by sediments as the seas advanced on the North American continent and the plants decayed and were compressed into coal. Oil and gas are harder to trace back to their origins, but represent marine life, much of it consisting of single-celled diatoms and algae, which were buried by mostly marine sediments. Later, the weight of the overlying sediments and rock compressed, changing the once-alive organisms into droplets of oil or a few gas particles. These then migrated through the sediments and rocks and were trapped by impermeable barriers. Over time, these became pools of oil or gas (this term is unlike the common notion of a pool, and instead refers to an area within a porous rock formation which collects oil, gas, and water.)

Oil and gas are found in rocks as old as the Cambrian Period over 500 million years in age. Some oil has even been found in older rocks, but these deposits probably didn't originate at that time. We are only talking about sedimentary rocks here—sandstone, limestone, dolomite, and shale. The current big oil and gas plays are in shale beds, which require large fracking jobs to recover the hydrocarbons.

We don't find oil and gas in igneous or metamorphic rocks which have been created by intense heat and pressure. However the Russians believe methane gas is created in the Mantle of Earth and migrates upward where it's trapped in near-surface reservoirs. The Russians have drilled a very deep well to try to penetrate into the Mantle of our planet and find gas. I don't believe they've found any gas to date.

Once an oil sand or deposit has been uncovered by erosion or faulting at the surface of the planet, the light hydrocarbon fractions quickly evaporate and only the heavy oil and tar are left. The La Brea Tar pits in L.A. are a rare and marvelous place to see an oil pool at the surface.

The United States has mountains on the east coast, the Appalachians; and mountains on the

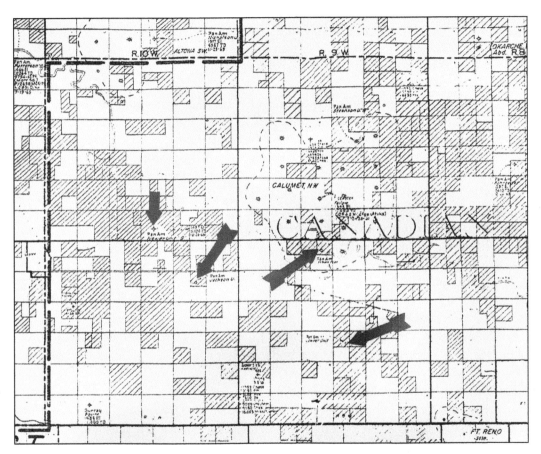

Pan Am's acreage holdings are shown as the cross hatched areas.

My first well recommendation in the area was a development well in Section 4-13N-9W, shown on the map as the Pan Am #1 Schweitzer. This was followed by the Pan Am #1 Louder in Section 22-13N-9W, the discovery well for the South Calumet Field. The Pan Am #1 Newer Unit B in Section 34-14N-10W dry hole exploratory well was not recommended by me, but was the second well I was taken to see logged when I first arrived at Pan Am. I located my second wildcat discovery two miles to the southeast in Section 11-13N-10W, the Pan Am #1 Jackson which was based on my interpretation of how the sands were deposited and updip from the #1 Newer Unit B. The field was named the SW Altoona Field.

west, the Rockies. The cores of these mountains are igneous and metamorphic rocks, and contain no oil. There are some intermountain basins within the mountains which do contain sediments bearing oil and gas. There are offshore basins off both the east and west coasts and these basins contain oil or gas and are the prime areas of exploration in recent years by the major oil companies.

The entire center of the U.S. is essentially without mountains and consists of a series of subsiding basins that received thousands of feet of sediments and limestone deposits from the Cambrian Period 542 million years ago up into the Pleistocene Epoch 2.6 million years ago. I think all of these states situated in the middle of the U.S. have produced oil and gas. Petroleum geologists regularly study these basins in their search for oil and gas.

These were primarily marine basins formed when the oceans flowed onto and off the continents. The history of the North American continent was simply the seas came, and the seas went out. This happened over and over again with rocks, especially in the Pennsylvanian, the Mississippian, and the Permian Periods showing this cyclical deposition—which are called cyclothems by geologists. The repeating sequence of rocks varies from tens of feet to several hundred feet in thickness, and the top layer is a thin limestone formed just offshore as the seas came in.

Sometimes these limestone beds grew into giant reefs—tens of feet to hundreds of feet in thickness. Backreef and offshore muds were deposited which were later compressed into shale. Organic life was high, especially in the back reef and lagoon areas. Sand is the next member of the cyclothem. The sands were deposited as beaches, bars, surge channels, and as dunes on the shores. The area inland was above the sea level and subjected to erosion that brought sands and mud off the land into the sea. Rising seas over the land sometimes preserved the stream deposits and channels.

You might ask, why did the basins form? In the mountains, and along the tectonic plate boundaries, faults and the stresses of moving blocks caused some areas to subside into a basin form. The interior of the continent is a different situation. Possibly deep rising heat currents from Earth's mantle caused some areas to bow upward. The areas between collected sediments form the rising lands. Once the process was started, the differential weight of thicker sediments in the basin versus the surrounding land area caused the process to continue.

In southern Oklahoma, the Wichita Mountains project above the surface to an elevation of about 700'. Just north of the mountains is the Anadarko Basin which is over 100 miles wide and 300 miles long, stretching from central Oklahoma into the Texas panhandle. Within a short distance, basement rocks of granite at the surface plunge to over 35,000' depth with sedimentary rocks above the basement rocks. The mountains went through several cycles of uplift and erosion. The sediments were all dumped into the subsiding Anadarko basin. At times, the oceans came over or around the Wichita Mountains into the basin, which contains mostly marine sediment deposits. At other times, the mountains were so high that the climate to the north was in a rain shadow area with very low precipitation, and red beds were deposited into the basin (red beds are formed in desert environments). Eroding granite sent granite washes miles north into the basin, again attesting to the height of the mountains at a given time.

The climate changed many times in a given region, and the types of life located there changed as a result— thus the source of the oil and gas varied, especially as the rapidity of burial was an important factor. If something dies or is killed and lies on the surface of the ground, it will be rapidly eaten, or it will oxidize and decay into virtually dust. If it is instead buried rapidly and placed in an anaerobic environment (meaning one free of oxygen), it has the chance of decaying and becoming oil and/or gas. Hence a subsiding basin is where you will find oil.

Well terminology

Drilling cost - All the expenses incurred for the drilling of the well. Includes cost of surface and intermediate casing.

Dry hole cost – Includes the drilling cost plus the plugging cost, running cement into the hole and cleanup of the property.

Producer cost or completion costs – Nearly half of the expense. Includes setting completion string of pipe, tanks and surface equipment, fracking, cleanup of the property, and fencing of the equipment.

Drilling a well

They say it takes over 100 people to drill an oil or gas well. It really takes more than that when you include all the pre-drilling activities. Then, after the well is drilled, many more people (and paper-work!) yet are required.

Let's say I am the originator of the prospect to drill a specific well. The idea comes from someplace — possibly a field study of an existing oil field. Maybe it comes from a rumor of a discovery in the area I am studying. Maybe it comes as a farmout from an oil company that would like to see a well drilled to prove up their lease holdings in the general area of the proposed well. All this may take weeks, months, or even years.

Next, we need a valid lease to drill on. A landman is hired and he visits the County Recorder's office and checks the records for existing and past leases. He also checks the chain of title to ascertain who currently has the legal title to the land.

The landman then approaches the title owner and negotiates a lease for the property — title, cost, royalty interest retained by the owner, and term of the lease. As originator, I pay these costs.

Commonly, the lease is drafted into a bank for payment within 30 days, which is supposed to give me time to check the title. In practice, we have already done that and the 30 days gives me the time to sell the project before I pay for the lease.

I find a company to buy the project and to drill the well as operator. We form an area of mutual interest for any more leases I may buy, or that the company may purchase in the area of interest. The company wants the right to drill offsets if the well is successful. I want an override interest on all the leases in the area of interest.

The company checks with the Corporation Commission for spacing of the unit. If it isn't spaced, then a landman or lawyer will file for spacing. If everyone in the unit has not leased or made a decision on joining, selling, or farming out, then a pooling will be required. This may take several months (so it's a good thing I bought a lease term of three years).

Finally, it's time to drill. A drilling contractor with the proper equipment to drill to the depth required is contracted. He subcontracts many other services and people needed to accomplish this task.

Bulldozers are required to build the location and dig a reserve pit. The pits are large and designed to contain oil, water, and mud if pressures cause an uncontrolled flow from the well. All pits are now lined to prevent fluids from sinking into the groundwater table.

The rig is moved in with steel pits for the mud, a shale shaker to collect the drilling samples, a generator for power, and water is purchased and piped or trucked in to the rig.

The proper lengths of drill pipe are delivered to the well. The lower part of the drill column is composed of heavy, thick-walled pipe that gives weight at the bottom of the drilling hole.

A drill bit company is contacted for the bits needed to drill the well. These may be tooth bits, carbide bits, or diamond bits.

The company has had their petroleum engineer design the drilling program, size of hole, and casing points to set intermediate strings of casing to prevent downhole high pressure zones from blowing out the well.

A mud company is contracted for a mud program and the various mud chemicals needed for a well in this specific area. Wells drill the fastest if the mud weight is slightly underbalanced against formation pressures. Water weighs eight pounds per gallon, mud weights can be double that in high-pressure zones. Most zones are lower pressure though, and nine to ten pound mud is adequate to hold formation pressures. We know from my previous studies the depth where the high pressure formations are. We want to set pipe above the high pressure zones and protect all the lower pressure zones up the hole.

The well is drilled and casing set in two or three different depths. The casings are hung one below the next, and the drilling hole inside the new string of casing gets smaller and smaller. Stabilizers are set on each string of casing to hold the pipe away from the wall of the hole and allow cement to completely surround the casing. (In the deep sea blowout in the Gulf Of Mexico, the scarcity of stabilizers was a contributing factor to the blowout occurring.)

Electric logs are run before each string of casing is set. Sometimes a small company will not run the uphole set of logs. Some electric logs can be run inside of casing.

Our well reaches total depth. The logs are run and the decision to set pipe and complete the well is made. Casing is run and cemented.

The drilling rig moves off the well and a completion rig is hired and set up over the cased hole.

The petroleum engineer decides on a completion program depending upon the type of rock formation and type of oil or gas reservoir encountered.

Another technical company comes out and completion guns are lowered into the hole and bullets fired through the casing into the formation.

Completion acid is delivered to the well and pumped into the perforations to clean out the cement. If it is a high-pressure, good permeability zone, the well may then flow and be completed.

Most times, a small frack job is undertaken to better connect the reservoir to the borehole of the well. Sand or glass beads may be pumped into the perforations to prop open fractures created by the fracking. The well then flows back into frack tanks which were just delivered to haul off the frack water. They used to just flow the frack water into the pits and let it evaporate.

The well is now completed. A Christmas tree of valves and gauges is installed on the well head. Tanks are set for the oil. A heater and separator may be required to separate oil and water and gas.

A pipeline may be laid to the well for the gas. This requires permits, digging equipment, payment to the landowner for damages, a contract for the gas, etc.

The well is completed and the lawyers prepare a title opinion comprehensively addressing the interests of all the landowners, royalty owners, participants, etc. Sometimes there are hundreds of people and entities involved in the ownership of a single oil or gas well.

Drilling a well is not a simple project and is like a puzzle with many adjoining pieces, all of which need to be snapped into place to achieve the overall commercial goal. I found it good to have partners who were lawyers, landmen, drilling operators, and accountants. During my oil days, I had a total of 47 different partners through the years. Most of our contracts were verbal followed with a simple one paragraph agreement simply stating that we were 50/50 partners in a particular deal, splitting all expenses and income. Rarely did we state what our individual duties were. It was understood, a landman was in charge of the leasing, the geologist wrote a report, and we both took part in the selling of the deal. It was a different time, fast moving, and with lots of action.

I put together one deal a month for eight years—which must be a record.

1971 dedication by the Oklahoma City Geological Society of a billboard showing the geology located underneath Oklahoma City.

The State Capitol building is located just to the right (west) of the oilwell in this photo.

♦ CHAPTER 4 ♦ Life in Oklahoma

A new life in a strange state…and a strange business!

It was June of 1964 and we were off to Oklahoma. Kay was pregnant and Brian was born in September. I had wanted to go back to the Juneau Icefield for a second season. The leader of the Icefield studies was Dr. Maynard Malcolm Miller who taught at Michigan State. He was willing to spring for an assistantship for me for the summer. Pan Am would not wait but insisted I move to Oklahoma in June—so my last chance to be a glaciologist disappeared. (I was beyond disappointed and this lost opportunity served as one more nail in the coffin of our marriage—it was such a tug of war for me between my professional aspirations and these unwelcome demands of family which had been foisted on me!)

Oklahoma was hot and dry in June and throughout the summer. We made a great deal on the purchase of our first home. A Texaco geologist had been transferred out of the area—and the company would not buy his house or help him get rid of it. It was a slow market at the time so we were able to purchase the house for just the $168 transfer fee of the mortgage. It was a small three bedroom one-story house with brick veneer situated at the end of a dead-end street—although we had never experienced living in an air-conditioned house nor the cost of cooling a house. The Catholic Church owned the adjoining 160 acres which was a marvelous place for the local children to play. To contribute to our new neighborhood, we put up a basketball hoop at the end of the street.

The Okies here talked funny—"y'all come back, ya' hear?" They were mostly southern Baptists. Our neighbor on one side was a lawyer from Oklahoma named John Patterson with not quite as much accent as his wife, Mary. The neighbor on the other side was from Texas, with a drawl so thick a non-native could barely understand. Our kids played together.

One day, the Texas neighbor from the other side came over. She was a little bit older than us, (late 30s) a tall, slightly overweight, and tough-seeming lady—a secretary by day, but a partier at night. Her husband was a "good ole boy"' truck driver who wore cowboy boots and a hat (how does one drive a truck wearing cowboy boots?). She was pretty well loaded and ranting up a storm about our kids. It seems our northern-raised kids with what we thought was no accent had influenced her kids. She was ranting about "y'all" and that now her kids were saying "youse guys!" We had a lot of drinks and resolved the problem. In reciprocity, I learned to use "y'all" and several months later tried it out on the District Committee when I was presenting a serious proposal. The Committee all fell on the floor laughing at me. (I subsequently resolved not to use either expression and rarely do to this day.)

Across the street was another neighbor who had the largest house in the immediate area, with four bedrooms. He was the manager of Reynolds Aluminum and was building a new house. Remember, the real estate market was slow at this time. We needed a larger house and I proposed buying the neighbor's house if we could sell our house. No one looked at our house. The neighbor said he would trade houses (for some additional dollars) and use our house for a rental until the market got better. It was a good deal so we agreed to that and purchased our second home. We closed the mortgage transfers and the neighbor's new house was almost ready for occupancy, but we were hot to move. They agreed to stay in our house and then move

again. So we got all our friends and neighbors, had a big block party, and moved furniture back and forth between the two houses. Some articles got moved several times, but a good time was had by all.

We lived there for three years and had a lot of fun times with our hard drinking, warm and gregarious neighbors. John Patterson was a pilot in the Air Rorce Reserve, and Jay Bond was in the Army Reserve. John flew regular missions delivering supplies to Japan. In civilian life, they were both lawyers, and each had significant influences upon my future life in the oil patch. We also became involved in the Catholic Church and made some good friends there too. This was during the term of Pope John, XXIII and the faith of American Catholics was being renewed in response to his leadership. It was also just after President Kennedy was assassinated and the Civil Rights movement was gaining steam. We were in an active discussion group with a dozen or more other Catholic families and also had a young idealistic priest heading our parish. Around this time, most of our group decided they wanted to be part of integrating part of the old city.

When the Oklahoma City Oil Field was developed in the 1920s, the oilmen built large and wonderful houses in one section of the city. These homes were on large lots, two stories — and were a mixture of architectural styles from bungalows to colonial revivals. Most of these early settlers had died by the 1960s. Many of these houses were subsequently purchased by blacks moving out of an area downtown which was being destroyed by the Pei Plan of urban renewal (the plan was to get rid of the blacks, bulldoze the area, and turn it into parks, government buildings, and large commercial developments. The downtown has still not recovered, but revitalization of the adjacent warehouse district has finally caused new growth downtown.)

Twelve families decided to buy homes in the new mixed area region. The houses were old, large, and cheap. They sold their expensive houses in suburbia for a good profit and integrated the old area. Father Narren followed them in the migration and was instrumental in forming a new parish with floating services held at people's homes. Kay and I had young children and liked our nearby schools, so we decided not to become integrationists at the time. Later, the bishop decided he didn't like the floating parish and ordered Father Nerran back to our parish. Father Nerran quit the church and moved to Sausalito, California. I don't know what happened to the rest of the group.

Kay and I remained in the regular parish church and continued to be active in the social events. I was still pissed over having to leave graduate school because she had gotten pregnant. I talked with my doctor about a vasectomy. He said because I was a Catholic, he would not do it without Kay's written consent. A year later I finally got her to agree to the operation which I subsequently had late one Friday afternoon. The next day was a church picnic in which I had agreed to cook hamburgers. I spent six or seven hours on my feet—the swelling and pain put me under an ice bag for the next two days. (I guess that was penance!)

At work, I had a great secretary whose services I shared with several of the other geologists. My mentor, John Mason, was having an affair with her best friend at the time. So many days after work I had a few drinks with John and Linda, and my secretary, Mary McClaren. I liked Mary—she was tall, good-looking, fun, but also serious. She always managed to slip my work into the mix before she finished some of the other geologists' reports. We were semi-romantic—kissed a little, but nothing else. She would never want to be responsible for breaking up a marriage. At this point, feeling frustrated and disgruntled, I was ready to leave Kay—but not yet. Mary was a career secretary with Pan American. When the company moved to Fort Worth, she was transferred to the general office in Tulsa and retired from there many years later. I occasionally talked with her through the years. She claimed she had a boyfriend, but never married to my knowledge.

◆ CHAPTER 5 ◆ Ramping up & meeting "The Wizard"

The Pan American Petroleum Company Exploration Manager was himself a geologist. The operating areas were split into Divisions with offices in different cities—Tulsa (main office), Oklahoma City, Liberal, Kansas, Fort Worth, New Orleans, Midland, etc. The Divisions were split into Districts (I was in the Oklahoma City District). The Districts had a District Superintendent who coordinated over the District Geologist, the District Landman, the District Petroleum Engineer, and the District Geophysicist.

Working at the bottom of the corporate heap, I was supervised by the ramrod of the office—a senior geologist, Herb Davis, whose job was to ensure that anything which went into or came out of the committee as a recommendation was accomplished. Herb was well liked by everyone in the office. He just naturally took command of any situation, with a suggestion, a joke, a pertinent story, etc. He quickly sorted out the necessary information management wanted to hear—in order to sell the project.

Herb was an Okie with a light drawl who was in his 40s, a few years older than me, about 6' tall and on the chubby side, with a round and heavy face. In always being able to take command of a room upon entering it, he would be serious or a cutup, whichever tactic he thought would be most effective for a given situation—he was just a natural promoter. He reported to the District Geologist, Howard Cotton, who had a clean desk with nothing on it while Herb's desk was strewn with the paperwork of every deal and ongoing project in the District. [Herb later left the company when they merged the Oklahoma City Division into the Fort Worth Division and most of us moved to Texas.]

There were 15 geologists in our district and a similar number in the Division, which gave me a broad learning arena given I was someone who had no courses in nor knowledge of the oil business when I arrived on the scene. Herb especially kept me out of a lot of trouble with Howard, who wanted to fire me, or at least rein me in. Howard's secretary would call Herb and say, "They're at it again—get in there quick!" Herb would then rush in with an emergency and whisk me out of Howard's office.

My memos and reports were never right in his eyes, so Howard would endlessly rewrite them. He particularly liked to rearrange the paragraphs. In those days, this meant the secretaries had to retype all these corrections, so sometimes my reports were typed and retyped five or six times. I came up with a plan and had Mary type the first report with several misspelled words. She wanted to correct them as any good secretary would do. I said, "No, I want Howard to find the mistakes." It worked, and Howard only had to correct the misspelled words. His comment was, "Tom is learning to write a good report, but he can't spell."

Howard also liked to recontour my maps, which incensed me. I told him that although my maps were not as pretty as his, at least my contours were on the right side of the wells! (I must confess that many geologists were able to draw beautiful contoured maps while mine often were not.)

Computer mapping is different and, in the early days of its use, produced very geometric looking maps. The computer looks at two points and regularly spaces the contours between the two wells in short straight lines. The next two wells require a different spacing. A busted point produces what looks like a volcano. At the edges of the mapped area the lines all flow together—like in a fault zone. Geologists make contours that curve and flow through an area as they think the contours and structure—with more data—would look like. Do the contours regularly space themselves between two points or is there a high or a low between those points? That's why it often takes more than one well drilled between two points to actually find an oil field following the initial research. I once drilled three wells in a row to find the oil field.

We came to work at 8:00 a.m., some of us climbed the 14 floors of stairs to our office for exercise, and work began with each of us in our offices. At 10:15 we collected for coffee—several times I was busily working on something and didn't want to go—however I was not allowed to skip coffee by my compatriots. We went to lunch at 12:00 (several of us brought a bag lunch), back to work at 1:00, at 2:15 coffee time again. At 4:00 p.m. we left work. Some people carpooled.

Usually I played chess during the lunch hour with Hal Brown. We never talked during the game—which was weird! He was difficult to really know in that he revealed little of himself and had a very closed-off demeanor. From a personal standpoint, he rarely smiled and did not seem to be a happy person. At the office, he never offered suggestions to me on projects I was assigned to, even those he had previously worked on. (Kay and I occasionally socialized with the Browns and, later, after Kay and I divorced and the Browns divorced, Hal and Kay got married!)

I spent the workday basically away from my office, reading all the old files, talking with any-one who had written a report on the area I was working, sometimes on the phone with ex-employees about a report they had once written. There was the occasional trip to Tulsa to the research lab. There was always a well drilling in my area that the company had an interest in. I went out to the well for the logging of the well, or to recommend a test, look at a potential pay zone, etc. I never asked for permission beforehand to go out to the wells, I just left word that's where I was headed. Then there was the paperwork on the 10 farmouts which I had to take to committee—tomorrow! Sometimes I had to get to the office by 7:00 a.m. so that I could get some additional work done. Often, I also stayed after work to finish up before the next day's committee meeting. The committee always wanted to see a map on the farmout area. (It takes a lot of time to make maps on the many potential oil-producing horizons in any given area). Sometimes I met with the company requesting the farmout and used copies of their maps, or I added my data to theirs to outline the play. (Since Pan Am had no intention of spending more money in the area for drilling, I was not giving away important information.)

Pan Am management only wanted to give away one drilling location at a time, which made it more difficult for another company to drill a possibly edge of the field well—without the oppor-tunity of drilling further offsets. I made verbal deals with several companies outlining how I would handle offset farmout requests. Give me several letters requesting farmouts on the adjoining acreage and I would submit them before other requests as soon as you give me the word. Often when the pay zone was cut on the initial farmout deal, I would receive a phone call requesting that I submit the next farmout request. I would then take the request to committee and recommend the farmout. The committee ruled that we had to see the electric logs or the completion reports on the well before they made the new farmout. The point is that the new request was always the first in line. If the well turned out to be a really good well, I always recommended to committee that Pan Am should drill the offset ourselves. The answer was no and the new farmout was made. (My ass was covered, and I had received no remuneration from the other company. Later when I had left Pan Am, my reward was a quick access to that company for a deal I was promoting.)

Other than Howard, higher management liked me as I got a lot done. By year two with Pan Am, I was getting more wells drilled than the rest of the office. (All were farmout deals from that dead area!) By year three, I was spending 50% of the company's District budget in drilling. Howard kept bugging me for never being in my office. I replied, "Do you want me to spend all my time in my office generating deals? I can easily spend the entire budget. What are the other 14 geologists going to do?" To slow me down, Howard kept thinking up ways to get me out of the office so that I could not go to committee every day and present all the farmouts. Initially, he sent me to Illinois with Doc Selk to run samples for a week and outline a different oil play (which wasn't even in our District!).

Next, he sent me to eastern Oklahoma with one of the Division geologists to do surface mapping in the Ouachita Mountains. I didn't even know the other geologist, but it turned out to be a great trip. We measured 50,000' of rock outcrops in the week. We drove every dirt road in the province and put maybe 10,000 miles on the company van. We drove down the roads at 20–30 miles an hour. Whenever we came to a good outcrop or saw a different-looking rock, we stopped mapped it, sampled it, and sped on. The weather was hot, dry, and dusty. We drank a lot of beer. At night, we stayed in the only motel in Broken Bow, Oklahoma. We visited the local liquor store to buy some wine to have with our nightly meal at the best restaurant in town. The year was 1967. Oklahoma was still basically a dry state and had only gone wet in 1959. My Oklahoma drinking friends liked it better when the state was dry. They told stories of calling their local bootlegger at any time of the day or night and requesting a wider variety of wines and liquors than one could obtain now in the state authorized liquor store. The Broken Bow store was really bad with virtually no wine, and for sure no good wine, only screw-tops vs. bottles with corks.

But then we stumbled onto a find!

We asked the liquor store owner if he could get any other wine ordered and delivered quickly. "Well, maybe by the end of the week. Wait a minute, I have a case of old dusty wine in the back room that some artist fellow ordered four or five years ago." He brought it out. It was a mixed case of very expensive French wines. We drooled! "How much do you want for the case? Oh, it's so old and dusty, I'll sell it to you for $2 per bottle" he replied, much to our wine-loving delight.

We couldn't believe our good fortune. We took our nightly bottle to the restaurant, which was allowed by Oklahoma law. They didn't own a corkscrew as the wines they dealt with all had twist-off caps. So we used the knife trick and splashed wine over the table. The next day we went to every store in Broken Bow looking for a corkscrew — not a one in the whole town, and there were no wine glasses for sale either!

Finally, I get a study to cut my teeth on!

Once back at the office, I didn't have to do any further work on the Ouchita project, since it was not my project. Then Howard came up with a new deal he thought might get me. He told me to examine a deeper horizon under the acreage I was farming out. This was the Hunton Formation on which one of the Division geologists, Ed Pitman, had recently completed an extensive study. The report did a great job of showing regional trends, but did not delineate specific drilling locations. I then collaborated with Ed and came up with a different idea on how to pinpoint porosity trends and erosional features which were dolomitized from the original Hunton Limestone formation. The Hunton was Silurian in age, and was a prolific producer of oil and gas in Oklahoma, Michigan, and other areas. The company Exploration Manager had cut his teeth in the District, discovering West Edmond Oil Field, and had made millions for the company. He looked at all the Hunton projects in great detail as it was his claim to fame.

With this new project, I had my work cut out for me, as I knew it would be highly scrutinized. The Hunton limestone was deposited in the Silurian ocean. It was later uplifted above sea level, and eroded by wave action and streams flowing over the land areas. When the seas came back as the basin subsided, the topography was covered with the Woodford shale of Devonian age — indicating a long period of slow subsidence. Mississippian formations of silty limestone covered the Woodford. My innovative idea was that the Woodford shale (previously an organic mud) was compacted by the overlying formations. Water was squeezed out and the resulting formation was compressed to 50% of its original thickness when it was deposited. To map this and accentuate the topography on the old Hunton surface, I simply doubled all the Woodford shale measured thicknesses. With this new map data and a lot of imagination in contouring the results, I produced wonderful maps showing a pattern of streams covering the old Hunton surface. This then helped us pick specific drilling prospects. I remapped the West Edmond Oil Field as I knew where this was going to end up. I read all the old reports and files, especially those written 20 years ago by the company Exploration Manager. Taking guts in hand, I proposed an offset prospect to the West Edmond Field.

Production from Limestone and Dolomite Carbonates

Limestone is formed of calcium carbonate, whereas dolomite is a calcium magnesium carbonate. Many of the mid-continent basins were filled with limestones, predominately lithified lime mud. Porosity develops in the seaward edge of reefs, in fracture zones, and occasionally in patches of Karst topography (think the sinkholes in Florida). West Edmond Field is situated in what is probably a Karst area. When it was drilled, it was reported that the drill bit dropped several inches to as much as five feet when the top of the Hunton Formation was encountered. In adjacent areas, the Hunton formation was dolomized by sub-aerial exposure and re-crystallization of part of the limestone. Dolomite has higher intercrystalline porosity than limestone. Many of the prolific wells in the area are from dolomite reservoirs.

Howard said, "This is really wonderful. Present this to District Committee." They approved it. He then told me, "I want you to take this to Division Committee, as you understand the implications of this project." Division liked it, especially John Mason. Howard then suggested to John, "Tom should take this to the General Office, it'll be a good experience for him." (I knew this was coming—everyone warned me that geologists had been fired on the spot if the Exploration Manager disagreed with their proposals.)

So, next, I was headed off from our office to Tulsa…to meet him: "the Wizard!"

Mind you, this regal-looking man in his 60s—"the Wizard"—took complete charge of every situation, always. His name was K.D. Soule and, in my day, his title was Exploration Manager—for the entire company, not just his division. The company was also ripe with countless stories of him firing people on the spot who dared to disagree with him in committee meetings. If someone offered a differing opinion from him, often as not, he'd also turn to those assembled and say, "Who brought this guy?" Needless to say, he was well aware of his power over his employees and he was duly in-timidating to one and all—a well-deserved reputation.

I was fully prepared and reread all the old reports, with a couple of pages xeroxed from an old report of his which he'd written 20 years before. John Mason introduced me and gave a very brief outline of my project. I was then put on the spot. The Wizard said he had a different interpretation of the regional area. I said, "No, you've forgotten your original interpretation in your report, Stan-13-04, in which you said thus and so." (John Mason was not-so-subtly shimmying down in his chair at this point.) I then whipped out the Xerox copy of the pertinent point. The project was approved on the spot and my fame spread far and wide in the company.

The next year I had a new project in a different area in which I wanted to buy 30,000 acres of leases. Back to the Wizard who approved 15,000 acres and four wildcats. They were only moderately successful and not worthy of a large drilling budget.

Back to the farmouts.

Several other companies drilled a few successful wells, but my best predictions [suggesting a large limestone reef with a Tonkawa sand on the basin side] proved to be wrong. The limestone was present, but had no porosity. The sandstone offshore bars were poorly developed, possibly because the basin subsided too rapidly in that area. In retrospect, I would have recommended the project again, maybe with a purchase of less acreage and different locations of the initial four wells.

By the way, K.D. Soule went on to serve as the president of AMOCO Europe.

Here's a funny incident I recall regarding good ole' Howard, the perennial burr under my saddle. One time, there was a stripper hired for an office beer fest, and of course all the fellas gathered around, eager to get a glimpse of her strutting her stuff. He was too short to see, so he stood up on a folding chair. Whether he was clumsy or just because he was a chubby man, the chair soon collapsed—causing him to fall and break his glasses, cutting his nose in the process.

This story quickly spread around the company and for days afterward everyone kept snickering and asking him, "Hey, Howard...how'd you hurt your nose?"

◆ CHAPTER 6 ◆   My fellow Pan Am Geologists

A group of a really different kind – so they say about all geologists!

Mike Maravitch was an old operations ramrod in our office. He sent people out to the drilling rigs, to check out pay zones, pick up electric logs, witness testing, or whatever. If something was needed by the geologists, he would get it. He had minimal geologic background but a hands-on knowledge of the overall business itself. He started out as a scout, spying on other companies' drilling rigs and leasing activities. He liked to share a lot of old stories which I enjoyed.

Frank Hine had the office next to mine. Frank was a quiet, slow-talking Texan who had grown up in the Texas Panhandle. When the company moved to Fort Worth, Frank and several of the other geologists bought homes in the southwest part of town. The area was wide open with no trees whatsoever, it was a very windswept locale. The tree line runs through Fort Worth, and then north, just west of Oklahoma City. Several of us bought homes in a northeast suburb of Fort Worth; my property had 42 small oak trees surrounding the house. I asked Frank why he didn't buy something with some trees around it. Frank replied in his slow Texas drawl, "Well, you know, some of us like to see 'em when they're comin'!"

He was an expert in drilling wells in highly faulted areas. He spent much of his career drilling in Eola Field. The pay horizons were 12,000'–15,000' under the ground. They were Simpson sands of Cambrian age standing on end. The sands varied from 10'–30' in thickness, with an oil column of 1000'–2000'. The trick was to hit a 10'–30' feet vertical zone three miles deep in the ground. The area was highly faulted and produced from several different fault blocks. Each well produced over 1,000,000 barrels of oil. Frank had wonderfully detailed maps. Maybe once a year, he proposed a new well as an offset or new well in a new block. The company always instantly approved his well sites. (Howard, I'm sure, couldn't understand his maps.) The well was carefully monitored during the drilling and carefully directed toward the target. Often the well missed the sand—maybe only a few feet away. Frank ran the samples or looked at cores, ran dipmeters, and recalculated where the pay zone was. The drilling rig circulated the mud, sometimes for several days until Frank determined where the pay sand was located. The well set a plug and a few hundred feet uphole or maybe further depending on how far away the target was, the well bore was diverted in a new direction. This is called whip-stocking the wellbore. The second time or possibly the third time, Frank would find the zone. These wells were drilled very slowly, often circulated for days while Frank figured out what to do. Rarely did he drill a dry hole. It was amazing to watch. Frank explained in his Texas drawl about apparent dip and thickness, how to interpret the dip meters and how he figured out the various fault blocks. There was no other area in the district with this complexity. (Again, I learned a lot about structural geology during this time.)

Bob Yedlowsky and his wife, Myrna, were from West Virginia and had the accent to prove it. I made a trip back to Ohio with him (another of Howard's schemes to get me out of the office) to review a big oil play happening at the time. Ohio had no rules governing the oil patch. Wells were being drilled everywhere—including in the grass strips between highways, next to houses and schools, even in a cemetery. It was unreal, and Pan Am chose to stay out of that play.

While back in Ohio, one of Bob's friends brought us a sample of bentonite clay found in a bog or pit. Bentonite is one of the chief chemicals (clays) used in drilling mud. Most of the bentonite used by the industry came from Wyoming at the time. So Bob and I thought this might be a great find and a business we could develop for ourselves. We took samples back to Oklahoma and had the Pan Am Research lab analyze the samples. It turned out there were too many impurities mixed into the clay for it to be economical. His wife, Myrna, was a trip and had a very definite opinions on many things! On her living room wall she had two artificial legs hung like pictures. It seemed they had belonged to her daddy who had lost his legs in a coal mining accident. Apparently, he had continued to mine coal on his artificial legs.

Myrna often suffered from excruciating pains in her head and neck—driving her to the brink of suicide. She was on heavy medication and virtually uncommunicative with the world at times. One day I passed their house and stopped in as I knew Bob was out of town and I had promised to check on Myrna. She was in the midst of suffering an attack, telling me she had pains shooting up the side of her head. She had received all kinds of tests from a variety of doctors—to no avail—possibly they thought it was her imagination. I had just come back from my dentist who had told me about one of his patients who had bad pain stemming from a poorly-aligned bite. I called him from Myrna's house and explained her problem. He said bring her right over and he would check her bite. It turned out that was the problem—with a simple fix—and she was cured. My dentist was a hero!

Bob was the absent-minded professor. He was always running out of gas as he never looked at the gas gauge, and Myrna did not drive, so he couldn't blame it on her. As a result, he carried a funnel and gas can in his car. Bob was also a chain smoker and three times (while I knew him) he blew himself up by lighting a cigarette while he was pouring gas—or afterward when gas had slopped onto his clothes. (He actually had a first rate intellect though and later left Pan Am to go to the West Virginia Geological Survey.)

Chuck Nunley was one of the district geologists, who had many years with the company. He was repeatedly passed over for raises and promotions and consequently had a bad attitude about the company. I received a raise or promotion every six months during the four years I worked for Pan Am. Why Chuck and I were friends amazed me, and I certainly didn't advertise my good fortune with Pan Am to him. He said if he had his career to do over he would talk like Henry Kissinger in a thick accent, just stroke his beard, say nothing of value and be recognized as an expert. "Yah, ve might drill right here! Dis a goot place."

Two of our good friends were Tony and Bonnie Benson. Tony was a handsome PhD, who was well-spoken, medium height with a full head of dark hair, while Bonnie was on the motherly-looking side—a somewhat oddly-matched pair it seemed to me. All of the wives of geologists in our

Here I am in surveying mode...

circle of friends were charmed by him (a few years later Tony and Bonnie were divorced). Although not too creative, he was chosen by management to be the deal presenter as he could pitch information put together by others. (As a rule, Howard Cotton, the District Geologist made the presentations — if they were brief and a sure thing; otherwise the geologist or Tony made the presentations.) Tony sat in on all the committee meetings and made the presentations to Division Committee, and much of the time to the Wizard himself at the General Headquarters.

Ron and Micheline McManus were also two of our good friends. Ron had his PhD and a good mind for the oil business. He was a hothead though and had several fights in bars and clubs with men who made a pass at his wife. They had a turbulent marriage. Ron and I had numerous discussions about corporations and many times he said he believed corporations were basically evil. I said "Why are you working here?" He didn't last long with the company and went back to teaching (a bit further on I'll tell you about our carpool experiences with them in Fort Worth).

Dave Lumsden and his wife were also part of our social circle. Dave was also a PhD and a quiet, very intelligent worker — not really suited to the industry. I think he also became a college professor. Kay and his wife Sandy were best friends.

Dave Graham and his wife were part of our little gang as well. The Grahams and the Cochranes were the only Catholics in our group. Dave was also a PhD — in fact, everyone hired during my four years with the company had their PhD., other than myself. (I guess John Mason had to have really seen something in me as you'll recall I only had a bachelor's degree and a couple years of graduate school!) The Grahams bought a house in a different part of the city. They then became good friends with their neighbors there who always seemed to have a party going each weekend. One weekend they invited the Graham's over for a wife-swapping party — that ended their new friendship in a hurry! After hearing the story, we kept teasing Dave that he should have given it a whirl!

◆ CHAPTER 7 ◆   Pulling up stakes for Fort Worth, Texas

In 1967, management made the decision to cut back domestic activities in Oklahoma and spend more money offshore and in foreign exploration. At this point, I was still the "fair-haired boy" and spending lots of the company's money—well, at least until my big project turned out not to be so big as we drilled four wildcats, all of which were poor producers or dry holes. The Oklahoma City District was merged into the Fort Worth District, and we all moved to Fort Worth, Texas. (Dallas would actually have been a better choice as it was more actively involved in the oil business.)

When this transfer to Fort Worth was arranged, Kay & I had never even visited the state before. Upon arriving, we couldn't help but notice the drawl the locals had here was worse than in Oklahoma!

I got pulled over by a local Texas cop because my car had an Oklahoma license plate. He was on a motorcycle and his uniform included jodhpurs and riding boots. He dismounted like he was getting off a horse and waddled over, asking me "…where I's from?" He then snorted, "Well, son, you know when you move here to Texas, you gotta have Texas plates on your car!" I told him I had just relocated to the state and he gave me a warning ticket, informing me I had to get a Texas license plate plus driver's license. When I went back to Oklahoma to check out a well, I immediately got stopped there for no other violation than having a Texas license plate. I was pissed and told the pleasant Oklahoma cop about my Texas experience, after which he laughed and then let me off. Turns out Okies and Texans don't like each other!

We bought our third house in Bedford, Texas, a dry suburb of Fort Worth. We found out that the next suburb, Hurst, Texas was "wet"—which meant we had to go there to buy beer or other booze. We could have bought a home over there—however we got a good deal on a house in foreclosure owned by the bank. It seemed the single mother, who had a 16-year-old son, was killed in an auto crash on the freeway in an area we called the "mix-master" with several intersecting and criss-crossing roads. The house was in poor shape with broken tiles, a couple of holes punched into the sheetrock by an angry kid, and it needed painting and roof repair. I took the house as is—the bank gave us $3000 credit to do the repairs plus applied that as the down payment. I spent maybe $600 for materials and put in all the labor required to fix the problems in the evenings and on weekends.

There were 42 small oak trees surrounding the house. Two were large enough to hang a hammock. I built a barbeque pit and we were living in style—beer, barbeque, and a hammock. What else does one need?

Sunday mornings, we drove several miles to the only Catholic church in the area. We got home, I grabbed a beer and proceeded to water some of the flowers and trees in the front yard. Our new Southern Baptist neighbor came over to say hello—and proceeded to tell me about the evils of drink and that I should at least not drink in the front yard. He was a refined-looking Texan in his middle 40s and delivered his spiel with a mild-mannered air. But it still pissed me off that he would greet a new neighbor with this judgmental type of "welcome." I have to admit, I repeatedly searched his garbage cans for booze bottles, but never found one. (You better believe if I had, he was going to get repaid with a lecture of mine in return.) Interestingly, Texas liquor stores sell more half pint flat bottles than any other state—guys can carry them in their breast coat pocket without a noticeable bulge.

I was in a carpool three days a week so Kay could have the car during weekdays when I was at work. Riding together were two other company geologists, a company secretary, and myself. Ron McMannus was one of the geologists—around my age with a good mind and a sense of humor. The other was an old geologist, who was also a Texan Southern Baptist. The secretary was a young, very proper, and nice-looking dark-haired gal, tastefully made up—and the wife of a Texan Baptist minister. Ron and I were shocked at their Southern Baptist prejudices against everyone—Catholics, northerners (Ron & I were both from back east), Mexicans, and especially blacks (niggers to them). Brenda was born in Fort Worth and thought Texas was the greatest and had no desire to see anything outside of the state. Pickup trucks were the norm and many sported rifle racks with guns. When I was driving, I would take an imaginary hand grenade out from under the seat, pull the pin with my teeth and toss the grenade in the truck next to us. After I had done it many times, they finally asked me what I was doing. Ron said, "Of course he's getting these redneck pickup trucks out of our way!" Then we went into a discussion about what is a redneck? We would pull up next to a really dark black man with heavy negroid features. I would say; "Wow! That's a really good-looking nigger!" (They always used that derogatory term). Ron would then correct me, "Call him a negro, not a nigger!" Brenda would say she thought that the black man was really ugly. Ron would jump on her and chastise her for only looking at others through a white person's eyes." What kind of a Christian was she?" We could steer them into any kind of conversation and end up with a bigoted reply. They couldn't help themselves—I don't know why they continued to ride with us!

It was so hot in Texas that I had an air-conditioner put in our VW microbus. It was a disaster and continually broke down. It was still under warranty, so back to the dealer it went. This malfunction went on for over 15 separate repairs so I finally gave up on having cool air in the vehicle. The courtesy van for the dealership was driven by a black guy I really liked. Often I was his only fare and I would sit in front with him. He was nervous about it and made me sit in back if there were any other customers. It seemed he had been a Civil Rights demonstrator and activist in the past. During one demonstration, he had both of his legs broken—reportedly by the cops. He had to keep his past a secret or be fired. We had good discussions about the movement as obviously I had no prejudice against blacks. When anyone else was there though, he would not engage in conversation. Sad but true.

Many of the stores we shopped in were poorly run by East Coast standards. I guess it was part of the Southern laid-back attitude. Their excuses were, "They forgot to restock," "Maybe we have it in the backroom," "Come back in a couple of days and we might get around to re-shelving the item," or "We used to carry that, the guy down the street might have it." At the conclusion of these frustrating exchanges, they invariably ended with "Y'all come back, ya hear!" I finally had enough and replied in frustration "I'll never be back—do YOU hear? You people don't understand business. I could bring in 100 New York Jews and we would clean out this state in a week!"

Our office was in the Commerce Bank building, which was an old 14-story brick building in the center of the downtown section of Fort Worth. The company gave us payroll checks drawn on Commerce Bank, and we also had our personal checking account there. We got paid on Friday afternoons at the end of the day. The bank was still open then, so we deposited our checks as we left work. The checks weren't credited until Monday morning. We wrote checks over the weekend, which were also deposited in the bank on Monday. The game the bank played was to credit the checks against our accounts which we had written over the weekend, and then credit our payroll check. Then they bounced the checks so they could charge us overdraft fees and bounced check charges. I was incensed and took my deposit slips to the bank showing the deposit one or two days before the bounced checks were written. Their excuse was that they had to see if the Pan Am check would clear. (Remember the payroll check was written on their bank and Pan Am ran millions of dollars through their multiple accounts with their institution.) It happened two months in a row. I switched to a different bank.

The office seemed real quiet. I actually spent more time there these days—working on a map, or running samples. We screwed around more and just chatted away with the secretaries. One day, the Fort Worth Vice President, whose office was two floors below mine, called and wanted to know who was throwing paper airplanes out the windows. Of course I told him I didn't know—"Maybe they came from the building across the street!"

I always used the women's restroom as there were only three women on the floor and the rest were all men. Restrooms in the old building alternated from a women's on one floor to a men's on the next floor. The ladies' was closer to me and I could spend a lot of time in there washing samples without having to instead lug them up to the next floor. I would yell to the girls that I needed to wash some samples—"Go if you need to as I will be in there for some time!" Howard, my boss would come looking for me. "Where's Tom?" he'd ask and one of the girls would reply, "Oh, he's in the girl's bathroom again! You want me to go in and get him?"

Since it was a quiet time with reduced budget and staff cutbacks, it was time for me to assess my future—stay with the company, go with a small company, or team up with one of my friends who had recently left the company? Herb Davis was a choice to consider, Ron Eddington was another choice, and there were others who knew me and my work.

Kay wanted me to stay with the company, she was focused on the fact it was a steady job and reliable salary, and she wanted to remain near her friends—all of whom were the wives of other Pan Am geologists. Pan Am was one of the better companies and easy to work for, although they had their rules which I loosely broke at every opportunity. Herb was no longer there as a buffer to save me from Howard's wrath. But the company tolerated me because of my high productivity. I always worked more than 40 hours per week so the fact that Howard couldn't find me at any given moment didn't mean that I wasn't working. (In fact, one of the final blows was when they asked for my office key to be returned so that I couldn't come in to work on Sundays.)

To succeed in the company, one had to aspire to management. If you didn't, then they looked at you differently, and just kept assigning you smaller and less-important projects—and you soon lost your interest and enthusiasm. If you had a special skill, one could migrate to the Division or the Research Center. There, an individual had more latitude in choosing their own work projects. But in District Operations, we (e.g. I) had no choice—we were supposed to focus our efforts exclusively on what was assigned. (You'll note I never followed these instructions and always worked on more than one project besides the one assigned.)

To aspire to management, one had to get the District Geologist, good ole' Howard Cotton, to let you present your deal to Committee. As mentioned, Tony Benson convinced Howard to let him help present a couple of complex deals (beyond Howard's expertise) to District and then Division Committee. With his personal polish, he did well and Howard came to depend upon him. I of course never asked Howard for anything—and also never asked him for permission beforehand to do something. I think Howard, knowing of my lack of background in Petroleum Geology courses, simply never thought I could ultimately make it in the company, he assumed I somehow got all my ideas from someone else. This was true of course as my investigative approach was to talk with everyone, read everything in the company files, and then put a new project together with my special twist—so the project became mine at that point.

When one became a manager, practically on any level, you no longer got to do what you (I) loved, which was studying the geology and carefully assembling all the pieces together for a new well to be drilled. If the well was unsuccessful, then the company usually just gave up on the project and went on to a new area. Since I never want to give up (unless I really busted the geologic interpretation), I always wanted to drill another well, but it was difficult to get the company to budget more monies for another well at that point. That's why I liked to promote programs with, say, four wildcat wells, rather than just one. If the money is budgeted for four wells, then if one or two are dry, they'd still drill the rest of the program.

One day in Fort Worth, I'm sitting in my office with Howard Cotton, the District Geologist, and Jake Warden who was the District Superintendent, as they were in the midst of a yearly review of my work. The phone rang and they told me to answer it; it was another ex-Pan Am geologist, Jim Richards. Turns out Jim was calling to offer me a job in Denver for one of the new oil fund companies. I said, "Thanks, Jim, but I'm not ready to leave the company at this time. Thanks for the offer." I told my reviewers of the proffered opportunity, hoping it would help in getting me a raise. And, indeed, they gave me yet another one.

Jim called me a couple of days later and we laughed about the timing of his call in the midst of my yearly review. I took a day off from work and flew to Denver and spent Friday and Saturday working on the terms of the employment being offered me. I also met their leader, we shook hands over the deal, and I said, "Put it in writing and I'll come work for you." However, it turned out he didn't want to put more than a very brief contract in writing. I declined the offer as something didn't seem quite right. I then warned Jim of my thoughts about the company and fears about where they were heading. He thought I was making a mistake—but later was screwed himself when the promoter split with the cash...and the company went under.

So I returned, lamentably, back to Pan Am. I hated Texas, both personally and work-wise. The new Fort Worth Division was a collection of about four other Divisions, operating in parts of four different states. The pace was slow and managers seemed to be everywhere. I could have probably gotten a transfer to Houston—which was an active Division—but I didn't like the climate there.

An opportunity knocks

And then, in 1968, Herb Davis, my old boss and compatriot (who'd kept me out of trouble with Howard), called and said, "Come back to Oklahoma City and join Bob Northcutt and me." Herb and his wife, Shirley, were both from Oklahoma and good ole' Okies. He was very outgoing, a little egotistical, and very ambitious. When Pan Am moved to Fort Worth, he had quit and opened this new business with Bob, an old friend. They'd been going for a year already at this point when he called me.

He said, "We'll pay your moving costs and subsidize you at $1,000 per month for a few months until we get some deals going. We'll split costs, profits, ORRIs and working interests one third each when we end the relationship, but keep the interests in the DNC Corporation while we're operating in the business." Wow, we had stock in the corporation, a partnership agreement, bylaws and officers, and operated like a real company. Color me impressed! Ok, Oklahoma, here we come.

As I mentioned, Herb was a great promoter, full of so many colorful stories and just as much B.S. He loved to sell a deal drawn bar-side with three colors of marking pens (which he always carried with him)—and on a paper bag no less. He could make a complicated prospect look simple and great with his three color method. Yours truly then had to produce the real map for the client and report for our company, Davis, Northcutt & Cochrane.

Herb knew the value of networking. We joined all the professional societies, became active in the local Oklahoma City Geological Society. Herb gave talks and we wrote short geologic papers which Herb presented locally and at an AAPG (American Association of Petroleum Geologists) regional convention. We directed a couple of small, wealthy independent oilmen's drilling programs. We sold many projects to a large and growing independent oil company before they hired their own staff.

Bob Northcutt and his wife, Annie, were the third part of our DNC Exploration Team. Bob had left Pan Am several years earlier and was working on his own deals. He was the son of a doctor from Ponca City, Oklahoma. Bob inherited some money, owned a beer bar in Ponca City, and slowly went through his inheritance. He took a long time to put a deal together as he was a "splitter" rather than a "lumper" (like I was). Lumpers look for the "big picture," whereas splitters can never see or finish anything as they're too buried in the details. Bob was good at setting up books, and doing legal work plus taxes. He kept track of everything. Bob was very helpful in reading my reports, suggesting changes and following up on the paperwork once a deal was sold.

Kay was reasonably happy as some money was projected to cover basic living expenses, and she had reconnected with her old church friends now that we were back in Oklahoma City. We sold the house in Texas for a profit and bought a nice place with a huge yard on Elmhurst Street. It was near the church, a different parish than our first one in OKC. It was the largest Catholic church in the area, and Kay got a job teaching at the church school. We became active in our new parish and ran the newcomers welcoming committee. We both had no clue how busy and hectic our lives would soon become.

Needless to say, I was glad to leave Texas. Upon returning to Oklahoma, I drove straight to the Motor Vehicles Bureau and got rid of my Texas license plate. (No more tickets.)

◆ CHAPTER 8 ◆   Leaving Pan Am & setting up shop—The DNC Exploration years

Another new chapter of life — wild n' crazy days!

Having formed the partnership of Davis, Northcutt & Cochrane, an additional operating company was also setup: DNC Exploration Corporation. I had to take a single office on the next floor for a few months until a larger space became available for our offices.

Our newly-expanded space had a reception area for our secretary, a conference room with a refrigerator and liquor cabinet (very essential in the promotion business), three offices for each of us, another office besides, plus a storage/file room. We sublet the empty office to a small drilling contractor who we used for shallow wells and to spud deeper wells.

Carol Holladay was exceedingly smart, efficient, and no doubt the best secretary I ever had. On top of that, she was a real looker: a short brunette who was a snappy dresser and always wore high heels—so when clients came to our office, I think it was often to see Carol. It would take her 45 minutes to apply her thick makeup, including false eyelashes, but she looked like a beautiful painted doll once done with her brushes, potions, and powders. We also put Lloyd Allen into the space—he was her boyfriend and, later, husband. Frankly, I think we business owners took Lloyd in to keep Carol close.

Lloyd was 6'3" and massively built, he looked like a football player—a rough-but-handsome man who was part Indian and the wild gambler type (it always tickled me to see him towering over her when they were side by side in the office). Lloyd always wanted to bet with you on something. One day, he said to me, "I'll bet you twenty bucks that Earl Pinney will walk in here in the next twenty minutes." I bit, as I had not heard that Earl was coming to town and he always called us before he came. Twenty minutes later, I'm looking at my watch for the sure win of a double sawbuck—when in walks Earl. I knew I had been set up, but I paid Lloyd the twenty bucks. No doubt Lloyd and Earl had colluded in this little scheme. Lloyd later got into quarter horses and lost a lot of money, but he had fun racing them. He actually should have bet on the other horses, as his never seemed to do well.

He needed a big loan to pay for some Mud Logging Trailers which he represented that he owned, and had sold to an oilman from South Korea. Lloyd always spent money as soon as he made it, so as a result he had little collateral for the loan. He banked in Ada, Oklahoma, in Indian Country. Lloyd whips into the bank president's office with his contract for the sale of the Mud Logging Trailers. The executive asked him what he had for collateral; although he knew Lloyd had always paid off his debts, he still wanted some collateral as is standard banking practice. Lloyd then whips out a pair of our secretary Carol's pink panties, showing him the diamond engagement ring wrapped up in the panties. It was a very large diamond and I'm sure Lloyd was still paying for it at the time. "Here's my collateral!" He got the loan. I'm not sure whether the bank president kept the panties, but I'll guess yes!

She had one failing though in that she had picked three losers before she met Lloyd. All three had beaten her, as obviously she was smarter than them. Lloyd, I predicted would be her fourth mistake, but I was proven wrong. He worshipped her, even when she was chewing him out. And thankfully he never laid a hand on her. After DNC split up, Carol virtually took over Lloyd's business and ran it successfully for many years. He finally cut back on his gambling, but eventually died of alcoholism. Carol reportedly nursed him to the end.

Shaking hands with an icon

Herb and Bob were the Okies and, between the three of us, we knew virtually everyone in the business. Herb had gone to OU (Oklahoma University) and Bob had gone to OSU (Oklahoma State University). Major rivals! Lots of oil money! I had farmed-out 100,000 acres so I knew all the small independents operating in the area. We also had an immediate "in" with any of the ex-Pan Am geologists — Pan Am was known as the industry's best training company.

We also became political on a local and state level and Herb became active in the AAPG (the American Association of Petroleum Geologists). He gave a talk or two, ran for committees, and became widely known in the industry as a result. We also helped promote and expand the Oklahoma City Geological Society oil library. I was an editor for three years of the *Shale Shaker* — the OCGS publication. Around this same time, a new society was forming — SIPES (the Society of Independent Professional Earth Scientists), and we became members of that as well. We also spent time at the Oklahoma University Geological department and Oklahoma Geological Survey. When an Earth Science teacher was drafted into the military, two of us local geologists stepped up and taught his course at the high school for the year. We supported Henry Bellman for governor, he was a fellow oilman. We opposed Senator Gary Hart of Colorado who wanted to place additional taxes on the oil industry.

The phones rang continuously, seems like calls came in from everywhere—New York City, Denver, San Francisco, Los Angeles, Houston, Midland, Tulsa, Calgary, etc. We traded jokes across the country, and it seemed there was always a new one to share. "Have you heard it? Yup, just did half an hour ago!"

We took our deals to Dallas, Tulsa, Amarillo, Houston, Denver, and elsewhere. We also cold-called continually and especially followed up on leads Herb developed at the industry conferences he frequently attended—one of these was to the well-known oilman, T. Boone Pickens. We were well known enough by this point so he agreed to a meeting with us to hear about Herb's (DNC's) "big deal." Pickens had several gas companies which were in the process of merging at the time.

We went to his office in Amarillo, which seemed modest for such a sizable operation as his. He was punctual and we were ushered into his office. A warm and friendly man, he sported a bolo tie which was commonplace, as well as cowboy boots with his suit, even more standard issue attire in those parts.

One of the things I liked about Pickens was his focus, he wanted to cut through the clutter and get right to the heart of the matter—in this case, our deal. He asked very thoughtful, probing questions, and then paused before saying to me, "That's a nice opportunity, sonny, but I only take BIG deals. You need to think Bigger!" We (I) certainly tried to follow that advice! We called his office a couple of times after that face-to-face meeting, but he was also highly selective about the regions of his deals. He started operating on a different level, buying, selling, and merging companies.

It was a wild time in those early years of our business with lots of money spent, lots of wells drilled, and we were part of all of it. Our bar bill was probably our biggest expense. Our refrigerator always had Bloody Mary makings, lots of vodka, scotch for me and Bob, and bourbon for Herb. Sometimes we even had food in there!

I opened the office every morning around 7:00 a.m., the secretary showed up after she dropped her kid at school, and Herb and Bob got in around 9:00 a.m. Clients rarely arrived before 10:30 or 11:00 a.m. By then, I had worked four hours on the next deals we were promoting. If I could get away, I went to the geologic library for electric logs or to get production data on wells around our projects.

I had resisted becoming an independent or working for a small company assuming I just wouldn't have the data to work with—in comparison with what was available inside the big oil companies. But it turned out we actually had access to more data now than I had at Pan American. Many of our deals involved farmouts from other large corporations—Mobil, Phillips, Anadarko, CONOCO, etc. So we often got to review their company data and their reports. We had the Geological Library, the Geological Survey, we subscribed to Petroleum Information, the Oil and Gas Journal, etc. I even bought a set of scout tickets (e.g well completion cards) from the widow of an old geologist whose tickets went all the way back to the earliest days in the oil patch.

The ones that got away

As our business flourished, we developed a good reputation in the industry and got involved in many more deals yet. If a major company wanted to farmout some acreage for a well, we were high on their list of people to show it to. We got a farmout from Chevron in downtown Los Angeles to work over a well, although the deal was killed by opposition from local residents. Wells are drilled and reworked in big cities. The rigs have giant mufflers on their engines to reduce noise. The rigs are completely enclosed with padded canvas, again to reduce noise levels for nearby residents.

We got a farmout from Shell Oil Co. in Cook Inlet, Alaska which required a $100

Dallas, Texas, July, 1970

The DNC team: (left to right) Bob Northcutt, Herb Davis, and yours truly, 1970.

million platform be constructed to drill and produce the oil. We sold half of it to one of our clients and obtained a verbal agreement from another client—but he had DeGolyer & MacNaughton run the economics. They decided it would require two platforms to drain the oil from the structure. This killed the economics—and the deal. (I'm not sure if the deal has ever been drilled to date.)

We missed a potential big one in the Paradox Basin in Utah. It was located at an elevation of 8700' and required a rig with supercharged engines to operate in that altitude. We had a farmout in Lake Erie for a gas well. Detailed geology indicated it was instead an oil prospect and we couldn't get a permit to drill for oil in the offshore waters.

A biggie that DIDN'T get away!

On a Thursday morning, we received a call from the local manager of Continental Oil. The voice on the other end of the line said to me, "I've got a great farmout for you if you can commence drilling today or tomorrow before the lease runs out at midnight." We checked our maps and confirmed it was indeed a great offset location Continental was going to participate in. He said the operator that owned half the section had failed to get a rig on location. "We can do that! We like challenges," Herb and I assured him.

We immediately called our drilling contractor, A.J. His rig was in Watonga, just six miles from the Continental location. A.J. says, "My rig just finished a well so it's available. I'll load my Cat and throw on a tin whistle (culvert) and be out there in half an hour."

He heads to the site which had been surveyed and staked with permits so that it was legal with the state. He throws in the culvert and cuts the fence. A farmer comes out and says, "You can't drill here. The lease has expired and I've given a top-lease to another company." By this time, A.J. has a spudder rig en route to the location and several trucks loading up with his drilling rig. He goes back to town and calls us with the news.

We call Continental Oil and they fax the lease to A.J. who then grabs his sheriff friend in Watonga and they go back to talk to the farmer. The sheriff tells the farmer we have the right to drill the well and, if he prevents us from access per the lease, he'll lose in a lawsuit that results.

It's now about 4:00 p.m. and A.J. moves the spudder truck mounted rig onto location and drills 30' and sets surface casing. By 6:00 p.m. the cement truck arrives and the well's 15" casing is cemented in place. The Cat operator has finished grading the location and started to dig a mudpit. In the morning, the mudpit is finished and fenced to keep out cattle. The drilling rig is being set up, the metal tanks and pits are set in place, the generator is hooked up for power. A.J. has acquired water from a half mile away from another local farmer. The waterline is being laid the half mile from the pond to the rig. In those days, 30′–40′ joints of two inch pipe with a rapid clamping mechanism were used for a waterline, and it took a crew of about five to lay these lines. A more modern system is now utilized with large spools of red plastic pipe rolled out across the landscape, especially as a leak in a plastic pipe can be easily fixed. Leaks in the metal pipe meant you had to change a couple of lengths of pipe, which happened with every well. The pipe was therefore laid in giant curves, which made it easy to switch joints.

By evening the well was drilling. We were busy in the office. We had to get the paperwork from Continental Oil, work out the details of the contract, pick up the title opinion from the other operator, get the other 50% to agree to our operation and agree to pay their half.

Most of all, we had to find a client to pay for half the well—approximately $500,000.

We called Earl in Denver, Colorado, who represented King Oil Funds. (We had previously given Earl maps of the area showing where the wells were located, and which ones were good wells. He could instantly check these maps and be able to reference much of what we knew from my studies.) He said his leader was out on his ranch with no phone out there, but he agreed it was a good deal for them and said he would drive out on Saturday to the ranch and obtain authorization to take the half interest—so we would know by Monday if we had sold the deal. If not, we would all hit the trail for a sales job. I'd work over the weekend and prepare maps and a report for a sales package—either for King Oil or someone else. Our secretary would come in early on Monday and type up the report. But everything worked out as we had put all the pieces together, unlike our friends at CONOCO, who had to take everything to committee for authorization. Incidentally, this one made a good well, we were heroes and we made a few dollars, even though our net interest turned out to be quite small. We did go to the top of CONOCO's list for farmout deals after that.

◆ CHAPTER 9 ◆   Deep drilling wells

Deep drilling in the 20,000'–30,000' depth was coming into play, not only offshore but in Oklahoma. Not to be left behind in this new trend, I began diligently studying all the deeper fields and looked for similar deeper-yet undrilled prospects. We got a farmout from Phillips to drill a fairly deep 17,000' offset to Aledo field.

Aledo field was an anticlinal feature at the edge of the deep Anadarko Basin. On maps it looked like two-thirds of a pie pushed up at the center as two or three faults radiated out from the center of the pie. Each fault block had two or three producing wells, with individual wells averaging about 20 BCFG each (billion cubic feet of gas). The water contacts in each fault block were at different structural levels. The wells cost more than $2 million dollars each to drill and complete. Five or six dry holes were drilled adjacent to the field in the water legs of the blocks. Phillips did not want to spend a million and a half for another dry hole. Hence the farmout to us. Productive wells in the Aledo Field produced six to ten times their cost. Of course, we wanted one of those! Even the four or five per cent that we could put as an override (ORRI) would be worth a lot of money.

The prospect location was on the down side of a seismic interpreted fault on Phillips leases, which would expire in a few months. New leases were going for over $1000 per acre and the spacing was 640 acres. Even if Phillips renewed the leases (if they were available), they would still need $1.5 million to drill the well, which might turn out to be a dry hole. The farmout was no cost to them and we were the suckers ready to take it (with other people's money, mind you).

Phillips let me review their seismic data in their offices. They kept bringing out different interpretations of the same seismic lines. Their geophysicists believed the fault was a big throw fault (e.g. having a large structural displacement) and the well pay horizon would be 2500' or deeper than the offset well. My interpretation of their seismic information was that the pay horizon would be 250' low and not 2500'. It turned out I was correct.

I got our drunken friend, Earl Pinney, to take a good piece of the deal. I briefed Earl on their seismic lines and map interpretation, as well as their pessimism. We headed over to the Phillips office—and, by the way, this was the most buttoned-down bunch of guys I knew at this particular company. I had our secretary's orange glasses with me. They brought out the seismic line—the one I liked. I said, "I have these special seismic glasses that let me see the structure. Here, Earl, put these on and you can see it!" I then carefully pointed out three beds on one side of the fault and three that I believed correlated on the other side of the fault. Earl was not a geophysicist, but could see my interpretation. We left with the deal and probably Phillips is still laughing over those clowns with the orange glasses! We drilled the well and my structural interpretation was correct, but the pay zone at the top of the Hunton Formation was missing. Possibly the fault cut the wellbore and displaced the pay zone. If our well had been drilled maybe as little as 100' from the spot we drilled, it would have hit the pay zone. We completed the well from an 11,000' zone and it made a fair well—but not the 20 BCFG we were looking for. As always, Mother Nature is complex!

Not to be abashed by this failure, we put a 15,000 acre farmout deal from Anadarko Gas Company in the deepest part of the basin. At this time, there were no wells in the area of any depth—so this was a real rank wildcat. We named it the "Cloud Chief Prospect" inspired by a nearby Indian town of less than 50 people. Seismic data indicated a large structure covering a township and a half. We picked a spot on the highest part of the structure, sold the prospect to a group of companies— each one with a different deal—different acreage cost, different net revenue leases delivered, different overriding interest to us, and some with a back-in working interest after payout.

My interpretation of the depth to our target, the Hunton Formation, was 20,000'. At the total depth of the well—20,000'—the potential pay zone was nowhere in sight. I estimated at least 5,000' deeper and possibly 10,000' deeper to the target. As a result, the well was plugged and nothing happened in that area for years, the leases just expired. Later, drilling on the prospect encountered a good gas sand at around 15,000' depth. Fracking made the sand commercial.

Another one we missed!

Since we were recognized as these great experts with deep basin drilling experience, we were nevertheless touted to Inland Gas Company back in Kentucky and West Virginia. Inland had drilled three deep wells on the western slope of the Appalachian Mountains into a deep trench known as the Rome Trough. The area had many shallow wells drilled on sample shows and minimal geology. There was no production from deep structures like those in the deep part of the Anadarko Basin in Oklahoma and the Texas Panhandle. We were hired to review the wells and give the gas company some geologic expertise on what these deep structures looked like and how to locate them.

Bob and I did the report and had Schlumberger Logging engineers from our area review the electric logs. The three of us flew back to Charleston, West Virginia and made our presentation and recommendations. That was a nice consulting fee. I'm not sure we really helped them find any deep production, but we learned a lot. That night we all got plastered at a local bar in the hotel. Bob claimed I drank 14 Black Russians (I think he must've exaggerated) — however somehow I became a guest singer with the band that night…and I don't know how to sing! Probably everyone in the bar left at that point — must admit I don't remember much of it!

Our final big play was located in southeast Colorado and found four wildcat discoveries, but these were all uneconomical for development. We were geologically correct, but unlucky on the completions. A larger frack might have helped. We had spent a lot of time, money, and mental activity on this play and basically were just tired and discouraged. Our major client, Federal Petroleum owned by Tom Fentem, decided to spend less money in future years and more time at the Oklahoma City County Club. We had spent a lot of his old oil money and had found him some production, but not to the level he desired.

One of the best potential oil deals we put him into was along the Wichita Mountain front in the Texas Panhandle. The well encountered a zone at 7000' depth that had over 100' of oil pay. We were ecstatic in that the average pay zone is only 10'–12' feet in thickness. The problem was the oil was very low gravity — almost like tar.

We could not produce the oil. We offered a big piece of the well to all the research labs of the major companies in the area. Maybe we could steam it out? Maybe we could flush it out with diesel or high gravity oil? Maybe we could inject gas into the zone? Nope — the pay is just too deep for the current technology. So, we plugged the well. After we split up our partnership, Herb sold much of his share of the production we had found and invested in the stock market with his portion of the proceeds.

I decided to never — or hardly ever — sell any of my production. Oil and gas prices fluctuate and may go higher! Redrills or increased density within the drilling unit may find new production. Over the years, I have made a lot of money from infill and redrills.

The pace of the business took a toll on our family lives though. Work, wine, women, and song got us. Bob and I had girlfriends and ended up divorced. Bob later remarried his ex-wife, Annie. I quit Catholicism and took up with atheist, Ann. Herb drank too much and was threatened by our divorces.

Bob and I continued to share an office and worked some deals together after Herb's departure, but we finally went our separate ways when he remarried his ex-wife. Herb, Bob and I all remained good friends though.

Heavy oil

Heavy oil occurs in many areas and presents a real challenge to recover. A famous area is the La Brea Tar Pits in downtown Los Angeles. These tar pits are breached oil fields where erosion has uncovered the oil beds. The lighter hydrocarbons have bled off into the atmosphere and the heavier tars remain. The La Brea Tar Pits collected water and animals who waded in to drink and bathe. They then became stuck in the tar and paleontologists are having a great time in recovering bones preserved by what is actually natural asphalt seeping up from the ground in this area for tens of thousands of years.

In Canada and Wyoming, there are vast deposits of oil shale at or near the surface of the earth. Major oil companies have tried various methods to extract the oil from the oil shales. Vast areas have been strip–mined as a result. These oil shales have been heated, steamed or cooked to extract the oil. Some areas have been burned to produce heat to drive generators for electricity. The impact on the land is worse than what we have seen in the strip mining for coal in West Virginia. Water has been piped in to steam or wash out the oil, thus producing another environmental impact. The cost is higher than drilling and oil prices need to warrant the added expense in order for these operations to ultimately be profitable. That said, the challenge remains as there are billions of barrels of oil contained in these rocks.

◆ CHAPTER 10 ◆   Plate tectonics—1960s-1970s: a change in geologic thinking

Big ideas rarely show up as an "ah-ha moment." The definitive proof of the idea may happen that way, but usually the idea, theory, or law develops slowly over time, with contributions by many sources.

Plate tectonics, the rifting apart of Pangea, was the big change in geologic thinking—but was not actually proven until the mid to late 1960s.

Alfred Wegener (1880-1930) came up with the concept of Continental Drift in 1912-1915—however his ideas were not accepted by the prominent geologists of the time. He died an early death in the Arctic while on a rescue mission. If he had lived a longer life, he may have further developed the theory.

This period early in the twentieth century was a time of big ideas which were embraced by an eager general public open to new thinking, new inventions, and exploration of the unknown, etc. Steamships, automobiles, electricity, airplanes, and other modern marvels were arriving on the scene, delighting many while startling older generations at the time. Agassiz proposed continental glaciation in 1840, Darwin published his work on evolution in 1859, both of which paved the way for grand ideas. Each of these concepts had been accumulating data and evidence for many years from many sources. However, it took a single mind like Agassiz or Darwin to put the evidence together and write the definitive paper(s).

The concept of Continental Drift languished on the back shelf. The data (evidence) kept accumulating though, waiting for a comprehensive paper to be written. Geologists became aware of much of this data. One of the persuasive papers written on the topic arrived in 1962, written by Hess detailing the mid-Atlantic ridge and spreading of the seafloor, and other papers soon followed. Once the theory of Plate Tectonics was presented, it was easily accepted by most geologists. Agassiz and Darwin had a more difficult time getting their ideas accepted in their day.

By the late 1960s when I was working in the oil business in Oklahoma, there were 15 geologists working in the office and nearly 100 geologists in other companies in the state. Surprisingly, there didn't seem to be much interest or discussion about the new plate tectonics news at the time.

I think I heard a talk on Hess's paper at the AMOCO Pan American Petroleum Research Center in Tulsa, however I don't recall any speakers discussed the topic at the Oklahoma City Geological Society meetings or at our evening geologic discussion group.

The evidence had been piling up for many years, even before Wegener introduced the idea. The 1906 San Francisco Earthquake should have sparked discussion on plate movements. Larson's work on the 15'– 22' horizontal displacement over 270 miles on the San Andreas Fault certainly proved that a large segment of the crust was moving. Studies showed the San Andreas Fault extended over a length of 838 miles with horizontal movement. However, it took many years of study to determine the extent of the movement.

During this period in the 1960s, we had many interesting topics to debate among our fellow geologists. We argued over studies presented in our Oklahoma City Geological Society Thursday evening discussion group which would later be considered as relating to plate tectonics. Bob Worthing showed us evidence for horizontal transform movement along the Wichita Mountain front which extends from south central Oklahoma into the Texas Panhandle. The vertical offset downward into the Anadarko Basin is over 30,000' displacement; if there was horizontal movement, it was ignored as insignificant. Lon Turk suggested the Nemaha Ridge extending from central Oklahoma north to central Kansas was a rift zone with transform movements. The so-called Nemaha Ridge consists of a series of anticlinal structures along a fault zone, basically down to the west. The regional fault does not appear to be one continuous fault but a series of single faults that terminate on both ends with a splay of faults. (We see this kind of faulting along the San Andreas Fault zone. Possibly the SAF splays are located at the end of a segment associated with a specific earthquake movement.)

Finally, in March 1969, the first real talk on the new Plate Tectonics was given (as I'd noted on an old discussion group agenda of ours). John Ramsey addressed us in reviewing papers written on the new concepts in global tectonics. I'm sure I was at the meeting, but I don't remember whether the concept itself was questioned or not. Many of us were just waiting for the concept "to be proven."

While at Pan Am, I had the good fortune of spending time with B.A. Curvin who was an original thinker. Bun, as we called him, had no specific project assigned to him. The company just left him alone, and would only occasionally ask him to look at a specific project. I loved his mind and would sneak into his office maybe once a week and spend some hours with him. From Bun I learned to look everywhere and read everything on a subject as well as anything that might be related. This included stumbling down all the blind alleys and dead ends—for there's surely a pony in that pile of information somewhere!

At one ongoing session, Bun was rifting apart the Mississippi Embayment from the Gulf of Mexico north to the faulting at Cairo, Illinois. The coal beds in Pennsylvania are the same age as the Morrow beds in Oklahoma which produce oil and gas. Both areas were thought to have been located in an equatorial location when the rocks were deposited. We had to drift and rotate the rocks in North America to get them into their current positions. Bun and I had no problem with this, if you are going to rift the continent apart, then there's no problem in floating the continent a couple/3000 miles north and then turning it a bit. I suspected Bun was correct in this theory.

I knew the Mississippi Embayment was different from other areas in the Mid-Continent. In 1960, when I was at Indiana University, we noticed we could see on the seismograph when storms hit the Gulf Coast. These micro-seisim's we saw in Indiana did not appear on records collected in Denver or on the east coast. We thought at the time the storms possibly affected the thick Mississippi Delta, causing these minor earthquakes. Bun and I thought the micro-seisim's were due to a different conductance in the underlying softer sediments. This was possibly another indication of the rifting apart of the North American continent. Maybe in 50 million years we'll know the answer.

I was thus prepared to grasp the theory of plate tectonics as fact. Today, geologists and much of the general public accept this concept of the movement of the continents as fact as well as an ongoing reality. Scientists are now wrestling with plate movements on other bodies in the solar system. The current thinking is there's no plate movement on our moon, none on Mercury, probably movement in the past on Mars, questionable activity on Venus, and they're undertaking further studies on other bodies in the galaxy.

Another current idea is that maybe a magnetic core is required to have plate movements occur. Perhaps a strong magnetic field is needed to shield the planet's surface from dangerous electromagnetic waves which may endanger life forms. The Earth is currently experiencing a weakening of its own magnetic field—possibly indicating a magnetic polar shift will occur in the near future. The Magnetic North Pole is wandering at a more rapid rate than in the past—possibly another indicator of a future pole shift.

Isn't it interesting how everything seems to be interconnected and how one bit of knowledge leads us from one place (e.g. idea) to another? I find it revealing to examine my connection to past events and people, which have led to how and why I hold my current beliefs. I don't really think I have ever had any truly original ideas, but like a sponge my mind has been absorbing droplets of information and concepts here and there which I have then, in turn, been able to articulate into some discussion or implement in some venture or other.

◆ CHAPTER 11 ◆ The making of "a wildcatter"

A development well is a well drilled offsetting a producing well, whereas a wildcat well is an exploratory well drilled outside and away from a producing area. A wildcat is drilled to find a new oil or gas field. The risk in each is considerably different. Offsetting a producing well has an 80% or higher chance of finding production. Wildcats have a much lower success ratio, likely only a 10% chance of success. Of those successful discoveries, maybe only 25% of them actually find significant commercial production.

Most wildcats are drilled or originated by major oil companies. When I worked for Pan American Petroleum, I caused six wildcats to be drilled. My first well found a very significant gas field. The second well had six wells developed around it. Two others were completed as small wells, and not offset for development. Two were dry holes. It never even occurred to me that I was proposing wildcats. That term was rarely used by anyone inside the company when proposing an exploratory well. We looked at the vast stretches of lease holdings of the company, and we proposed wells to prove up the acreage. The proposed wells were located on trend with existing production (sometimes many miles away—but still on trend). In areas having no production, these wildcats are called simply "stratigraphic tests." Even big companies don't want to call an exploratory test a "wildcat."

As a consultant or small company geologist, I myself never used the term "wildcat," especially as everyone knew wildcats have high risk. We wanted to acquire and draw in investors, so we played down the risks accordingly. Geologists and geologic studies are inherently risk reducers. There is always the target formation of the wildcat, but to reduce the risk, we show maps on secondary targets. If one can show that three or four zones happen to all cross your exploratory well location, then the prospect becomes very saleable to outsiders. Several wildcats I drilled missed the major target, but found significant production from a secondary zone.

The ultimate goal and ego stroke is to discover an oil or gas field, e.g. I had an educated guess and I was proven right. Making money on the project is great, but the satisfaction and thrill of finding a new field was what kept me going.

The following anecdote is my fantasy of how I became a wildcatter

I wonder if I was struck by lightning while on the mountain?

In the beginning, a long time ago, I climbed a mountain—just like the bear, to see what I could see. While sitting atop the mountain and contemplating my future, a small squall arrived and a rainstorm began. Once it ceased, a rainbow then suddenly appeared. I looked closely to see where it touched the ground. Everyone knows: where the rainbow touches the ground, there's a pot of black gold buried. I whipped out my map and made an 'X' on the map where the rainbow touched the ground.

I came down off the mountain and wrote up a description of my experience and vision. I showed the map and report to several of my friends and investors. " I have a vision and I know where the pot of gold is!" My enthusiasm succeeds in instilling my vision into the investors' minds. They rush out and buy some leases around the 'X' spot to drill a wildcat well. Soon the vision is theirs, and it of course morphs from my original vision. They decide to drill a half mile from the 'X' and the well is drilled. It makes a small well, but doesn't find the pot of gold as I promised.

I climb up the mountain again and take another look down at the 'X' spot. I think I hear laughter in the bushes from the Leprechauns. They're singing a taunt, "We moved the pot of gold so you couldn't find it, but we dropped a few coins from the pot. See if you can see where we put them!" I see no sign from the heavens though and now don't know where to look.

Each day I climbed the mountain again and looked over the landscape for where the pot of gold could be hidden. Finally one day, there's a new rainbow, bigger and brighter than the previous one. I mark the 'X' on my map at its end location and hurry down the mountain. I tell my investors of my new vision and say, "I know now where the pot of gold's really located!"

Once again, they believe me and buy some leases, and this time stake the well precisely at my 'X' location. The well is drilled and makes a fair well, but still doesn't find the pot of gold we were seeking.

I re-climb the mountain and look once again over the landscape. I hear the Leprechauns laughing and singing once more, "You almost found it that time. We barely had time to dig up the pot of gold, but we spilled some coins again. Give us more warning next time!" Again I look over the landscape and no sign from the gods. "I guess I will come back tomorrow," I say to myself as I scramble back down the mountain.

I climb the mountain each day and there's still no further sign from the sky. Finally, after many months, another rainbow appears, the most glorious of them all by comparison. "Surely this is THE one which reveals the location of the pot of gold!" I mark the 'X' on the map where the rainbow ends, this time bigger and bolder than before. I come down off the mountain once again, write up my report, and tell some new investors of my vision. They buy into it and acquire some leases to drill a wildcat. Being new investors, they move off the 'X' and drill the well. It's a dry hole. I can't believe it, especially that they didn't drill as instructed at my specified location.

I then convince them to drill another well. Again they move the location just before drilling the well. Again, it's a dry hole. So much for that set of investors!

Since a real wildcatter never gives up, I decide to re-acquire the leases and drill a third well looking for the pot of gold. I revisit the mountain and look at the 'X' location. I hear the Leprechauns laughing yet again, "You can never find it! We left it where we placed it and you looked in a different spot. I guess we don't have to worry about you again. We're going on vacation. Ha-ha on you."

I get some new investors yet again and we start drilling at the original 'X' location. The Leprechauns don't show up and we find the pot of gold—not as big as I'd hoped, but, after all, they'd already spilled out some of their coins on previous wells.

So I'm here to report there IS a pot of gold at the end of the rainbow, you just must be persistent in finding it! I am thus hooked forever in looking for this hidden treasure which lies buried deep within the earth. The gods have ordained it and, like Sisyphus, I must trudge up the mountain each day and search far and wide for the next new rainbow.

"Can I interest you in investing in a [that verboten word, "wildcat"] well….ahem, I mean a low-risk exploratory well? It's just sitting there waiting for us to find it!"

◆ CHAPTER 12 ◆   Life changes again — a new chapter begins...with Ann

Be careful who you meet for a drink!

It was a day that changed my life—May 5th, 1972—the day I met Ann. Among other significant developments that followed, meeting her led to the end of the Davis, Northcutt & Cochrane partnership, as well as to the end of the other partnership in my life—which meant divorcing Kay. Thus began a new chapter of my life with new partnerships, new businesses, finding The Sea Ranch, raising 10 children through their teenage years (mine, hers & theirs). Our paths originally intersected because my partner, Bob Northcutt, told me on several occasions I needed to meet Ann Kernal who was a secretary at the MacAfee Taft law firm, our corporate lawyers. At the time, she was the secretary to John Patterson who dealt with our account and who had been my neighbor when Kay and I first moved to Oklahoma in 1964. Bob was always the point person for dealing with the law firm as his cousin was Bob Taft. Although Ann had dealt with our account for a couple of years, I had never talked to her, but I did occasionally talk with Patterson. John also told me I needed to meet his looker of a secretary.

Bob arranged for Ann to meet us at noon for lunch at our favorite watering hole, "Over the Counter," which was a small bar/restaurant in the First National Bank Building in downtown Oklahoma City. Bob and I had a couple of Bloody Mary's and did some of our usual plotting. Time passed, and no Ann. Finally, about 12:30 she arrives, a lovely blonde with wavy hair and blue eyes—but there was immediately an intensity about her too. She was looking very glum and somewhat tearful so I said, "What's your problem?" Ann replied, "I was just turned down by a new shrink I had an appointment with. He wouldn't take me as a patient because he said I was suicidal!" I was taken aback but replied, "Hell, what you need is a drink and a kiss." And I proceeded to give her a kiss. She was startled and so was I as this was not how I usually behaved when first being introduced to a lady.

We then had several drinks and she came out of her funk. We proceeded to tell each other the stories of our lives and the afternoon wore on. Bob left as he had to get some mail out that afternoon, and we were then alone. We never went back to either of our offices. I don't know where we went—but the conversation continued.

At midnight, we were sitting atop a large dirt mound at a construction site in southwest Oklahoma City—a spot neither of us had ever visited before. It was a bright, starry night and our bantering conversation continued. Ann pegged me as a middle-class, somewhat boring geologist type. I pegged her as a 60s hippy. We did not have sex. I don't recall that we even kissed other than that initial kiss I gave her in the bar upon greeting her. At 1:30 a.m., we decided we needed to part company and get some sleep—especially as we each had to get back to work in the morning.

We met each noon for lunch the rest of the week, and after work each day for a drink or two. Ann had her roommate friend, Sandy, check me out. Ann had three children: James, who was living in McPhersen, Kansas with his grandmother, her daughter Shawn who was 15 and living with Ann (plus in the process of discovering boys and sex) and, Hadyn who was five and just starting school. Ann had been married twice, once to her high school boyfriend, producing James and Shawn; and her second husband was a man named Harry Kernal, who was one-half Indian. Both Ann and Harry had high I.Q.'s so it was no surprise their daughter, Hadyn, was super smart—although she also had a dark and somewhat brooding manner. (When a teen later, she went through a 'goth' phase which included always wearing black clothes, she even painted her bedroom black. While I was pleased to learn she later became a college professor, I've subsequently lost all track of her.)

Two weeks later, I had to check out a drilling well over the weekend. I invited Ann to go along with me. I rented a small cottage on Canton Lake near the drilling well. Sandy arranged to look after Hadyn and Shawn.

I arrived early Saturday morning to pick Ann up at her apartment. She was exceedingly nervous and so was I. We both knew we had to deal with the big "S" thing that weekend. She told me, "I've got a few things I want to take with us." I replied, "No problem, let me help you load them." She proceeded to bring out a suitcase, a small bag, a stereo with two large speakers, a large stack of (78) records, several books, a sleeping bag (I don't know why?), a cooler with lots of food and drink, a tablecloth, wine glasses, and I don't remember what else—good thing I had a large station wagon!

The old rustic cabin was secluded and overlooked a glassy lake, surrounded by tall trees. It was autumn, so no one was around, providing us with lots of privacy.

Our weekend together was great. I was introduced to many wonderful musicians—including Cat Stevens—who became my favorite musician. Ann tried to claim him, but I always insisted he was mine. She read me passages from some of her favorite books. She was an intense reader and could quote all kinds of people and subjects. Ann developed a headache, decided to take a bath, stripped her clothes off—and we took turns reading "The Little Prince" to each other (!).

Later that first day, I was on business after all, so I checked out the drilling well. The big "S" thing loomed in both our minds. As the day wore on, we tried to discuss it, but to no avail. Later on, we kissed. I fumbled with a button or two. Ann jumped up and said, "No one is going to take my fucking clothes off! Take off your own!" She then rapidly pulled off her clothes, I took off mine and we did it. She cried! I cried! We were in love!

We tried to deal with our emotions and this mess of a situation rationally. We then both tried to stop seeing each other, a resolve which would last on and off for a couple of days, and then we would go through it all over again. Even if my heart wasn't in my marriage at this point, I was still a Catholic, married to Kay, and the father of four children. Ann was a third-generation atheist, twice divorced, with three children. I'm easygoing. Ann was confrontive, combative, bi-polar plus off-the-wall overall. She suffered from significant mood swings and depression, sometimes within short spans of time. She attempted suicide many times, but not seriously—probably only as a means of gaining attention, or as a sort of mental escape.

With all these above-mentioned problems, a normal person would have quickly exited the scene—so, what was my attraction? Probably hormones to some extent, but that was not really the main driver. Ann and I were able to talk about everything and seemed to be on the same wavelength—although our discussions were always intense. Like most men, I have difficulty in really examining my emotions and deep beliefs.

It's a long way from a staunch Catholic to an Atheist. Through the years, we examined most of the major religions, read many of the world's noted philosophers, studied death and dying, the possible hereafter, heaven and hell, moral beliefs of various religions and societies, and you name it—we discussed it.

On top of all that, we had families to raise, a business to run, lives to lead, and sports—particularly football—to watch. We all loved the Olympic games. We purchased a big screen TV and the children all gathered 'round in our bedroom sitting area to watch this global competition by the world's most stellar athletes so we could interact together as a family. We usually ate our meals in the bedroom in front of the TV or sometimes by the swimming pool. The kitchen and dining room tables were probably used more by the children as places they could spread their books out to do their homework.

Working together with Ann

Along with her working history as a legal secretary, Ann had experience preparing tax forms for a CPA, and had also previously been a secretary in a farm implement business.

Ann had solid business instincts and, for example, at one point she had managed the rental of 100 storage trailers for the accountant/owner. When she took over the business, the accountant had no idea where the trailers were located and how long some of the clients had been using them. She ran down where all the trailers were being used and got the clients current on their rental fees. She could tell her boss from memory where every trailer was at a given time. One would-be new client she turned down when he asked to rent a unit. He then saw her boss who told her to rent him the unit. Ann informed him that she thought the man was a crook wanting to use the trailer for some possibly illegal business. She was right—her boss ignored her advice, rented him the unit, his check bounced, and the trailer was never found.

As a result of her good head for business she helped me formulate some of the more complex deals. She never objected to my handshake deals with partners and clients. Turned out her father believed in handshake deals too. He was a wheat farmer in McPherson, Kansas, but he owned 22 rental houses (very low-end), and other pieces of property there. The best income property he owned in town was a used car lot. He gave one rental house to Ann, which we had to clean out and remodel in between renters. Ann later sold it for $16,000, which she put into our oil business.

Ann's father died just before we met, so I never got to know the man. But Ann, her mother and her two brothers told endless stories of his keen business acumen. The family had a large farm of nearly 1000 acres which brought in a good income from the wheat they produced. However, Earl Mathes was a hard driver of his family as well as his employees. The old farmhouse they lived in didn't even have a modern bathroom. They used an outhouse in the backyard and heated water on the kitchen stove for their baths.

About Ann's last year in high school, Earl did buy a wonderful old Victorian in town. However, he did not like living in town and often stayed at the old homestead. Brother Jack inherited the farmstead and most of the land when he passed. Jack then remodeled the old place and it looked pretty good in due course. He also bought new farming equipment with air-conditioned tractors and put on his ear phones so he could listen to country n' western while he was plowing the fields. (Father Earl had never in his life bought a new piece of equipment. Everything was held together with used bailing wire, but he somehow kept the machines working.) With this modern new equipment Jack and his wife farmed the entire 960 acres with only one hired man during harvest season. (Earl had always arranged for Boyd and Ann to come home for harvest time so he didn't have to pay hired hands.)

Jack bought a big RV and, while the wheat was growing, they travelled around in their motorhome. They chatted on their CB's to all sorts of folks located all over and would meet up with other strangers for a beer someplace or agree to camp at the same campgrounds with people they met on their short-wave radio. "Come on! Big Bertha and Farmer Jack here. Any smokies out there?"

Jack was a solid-looking and very strong man, and you would immediately say he was from the Midwest. I forget his wife's name, as both Ann and brother Boyd referred to her as "piggy." She was indeed overweight, with squinty eyes, and very thick glasses. She had a high rasping laugh. But they loved each other and their family farm…and stayed married for life!

Brother Boyd was the runt of the litter in size, but not in brainpower. He went to college and got a business degree, after which he was employed by Motorola in Phoenix, and rose in management, eventually supervising a plant with over 100 employees.

Boyd and Ann were exceedingly competitive with each other while growing up. Later, when he moved to Phoenix, we would occassionally visit him for a couple days. While there, out came the games—foosball, billiards, and cards. Boyd could not stand to lose, especially to Ann. By the time we left, his ulcer was in full uproar and he was clutching his stomach.

He was the only one in the family who wasn't an atheist. Ann's mother went to church as her lady friends were all Christians, but she did not seem to believe any of it. However, Boyd carried a bible and used it to date women. He was such an uptight, hard, and opinionated person that his relationships didn't last long. He would date a girl for two or three months and then propose to her. They wouldn't have had sex before they were married. Almost a year to the day later, he would divorce the woman — I think that his marriage tally was around 10, but one of them he married twice. With each new marriage he also bought a new house. I think in the divorce settlement, he would always give the house to the wife being divorced, especially as he was always the one who initiated the split. Obviously, he was looking for something in a woman that he sadly never found.

My son, Brian, with Ann's daughter, Hadyn, standing on the pipe rack at a job site.

◆ CHAPTER 13 ◆   Oil patch characters

Truth be told, I met a lot of strange characters in the oil patch! In boom times, everyone thinks he can find oil right up to and including bartenders and used car salesmen—they buy leases everywhere and promote the lease as covering millions of barrels of oil. I've seen them all. A furniture store owner mentions in the TV ad for his store that he has a hot lease. You wouldn't believe it… oil fever spread to seemingly anyone and everyone in an on-fire oil region—like a disease that was highly contagious!

Once I was staking a well based on my geological research and a stranger shows up telling us we're drilling in the wrong place. He whips out his copper divining rod and shows me where the oil is. We should move over 100', he says. I drill the well at my location and it makes a small well. Hmmmm, maybe he was right?

The Drilling Crew

Wells drill 24 hours per day, with two or three shifts. The well site may be located 100 miles away from the homes of these crews, so they have to make the long drive roundtrip each day. It's tough and dirty work, especially when it's raining or the drill string is being removed for a new bit. A little gas in the mud makes each joint connection squirt mud.

The rig is manned by the "pusher" who is the drilling superintendent. He is on-call or at the rig literally 24 hours per day. If something's needed, or broken, or a pay zone is about to be drilled, he's on-location. He calls for all the outside services, when required.

The well is drilled by the "driller" who controls the drilling operation itself. He changes the weight on the drill bit by releasing the brake, he may also change the speed of rotation, and, overall, he directs the rest of his crew of "roughnecks." It takes one man in the rig tower when the pipe is going in or coming out of the hole as his job is to throw the 90' feet string of heavy pipe into the rack. The driller is operating the machinery to pull the string. It takes two crew members to operate the huge powered clamps which screw or unscrew the pipe together. They all work together in a very rapid, fluid action and the pipe quickly emerges out of the hole.

Amongst themselves, these guys all talk like a bunch of "good ole' boys." Even though one of the crew was a high school teacher, his calling changed when he learned there was more money in being a "roughneck!" One of them reads poetry during the long nights. A couple of them are "recovered" alcoholics (we hope). The flunky of the group has to go and catch the samples every 10' for the geologist, who might show up at any time. Another one does most of the mud mixing. The driller will call for raising the mud weight as the pusher has ordered him to have the mud up to 10.1 pounds per gallon by 9500' depth.

The drilling is orchestrated from a "dog house" located on the drill floor. The crew hangs out there when they're not occupied with other activities. There are no bunks, just the drilling controls, electronic recorders, a hot fire to keep warm by, and metal benches which are really lockers containing equipment. The crew usually brings their lunches with them and typically a flask of coffee. Mind you, there is always a coffee pot going in the pusher's trailer and the mudlogger's trailer as well as in the doghouse. The nauseous smell of diesel seems to permeate everywhere and most of us just weren't very hungry as a result, in spite of the hard work and long hours on the job site.

The Mud Logger & Wellsite Geologist

A separate company is hired to monitor the mud and samples for shows of oil and gas. The "mudlogger" is generally a geologist with lots of sample running experience. His trailer has a bank of electronic equipment. He hooks up to the rig generator, although he may have a small generator of his own. He has a sink, a stove, and a bunk bed for two. Port-a-pottys are located outside. The mudlogger sets up his equipment part way through the drilling operation and remains on the well throughout the remainder of the drilling. He works 24 hours per day and may be on-location for two or three weeks, or even longer on deep wells. He slips into town for supplies from time to time, maybe to take a shower and get a meal or drink at the local café.

When I came out to the well to see a particular zone, I would stay with the mudlogger. Often, the mudlogger catches his own samples as he feels he can get a better representative sample than the sloppy crew sample. It's interesting to see what books he's reading. We trade comments about the goings-ons and often share life stories as well. There's lots of downtime in the oil patch while we're waiting for something to happen.

The pusher usually has his own trailer on-location. If it's a well-maintained rig in high demand, he may have a very nice two or three-bedroom trailer. Company men may stay there for a night or two. We all gather there to look at the electric logs and make the decisions to set pipe or plug the well.

The mud company will often have a storage trailer on-site for the many bags of mud. Sometimes the mud is just left on pallets.

The wells are logged from truck-mounted logging trailers driven onto location at the time of logging. They usually have a bunk for the logging engineer, who may end up on-location for more than just a day. If the electric logs don't go down the wellbore because samples or the side of the hole is plugging the well, a trip with the drillbit will have to happen, and will take several hours to clean out the hole. During this time, everyone waits with anticipation—shall we take a nap or shall we go into town for food?

Other hired hands

Pumpers

After the well is completed, you must have someone monitor the well's production. Our office is 100 miles away from the well. My partner, Bob Marple is a petroleum engineer, so he designed and setup all the equipment needed to produce the well. We would hire a local person who lives near the well to be the "pumper." His job is to check the well daily to see that everything is working properly. If we're lucky, the local landowner or his son may be a pumper, or we can train them to be a pumper. The landowner wants the well to work properly as he receives compensation from it each month. Usually pumpers take care of several wells as they don't receive much money per well, but then they don't usually have to spend more than 15 minutes per day at each one.

Once the tank is full of oil, the pumper calls for an oil pickup. If the water tank is full, it needs to be picked up too. If the pump breaks a belt or something, he may repair it, or he may call to have it fixed.

He also watches the well—especially as there are people who show up to a working well to steal oil. One time I was sitting on a well and had to make several trips out to it each day, so at one point as I was passing one of our previous wells I decided to stop and check in on it. It was a gas well and we were not producing any distillate, which is normal for gas wells. I noticed lots of truck tracks into the well site and checked the oil tank. I could hear oil (distillate) dropping into the tank, so I knew the well should be selling oil. The valve was properly locked, to which our pumper had a key. So too did the oil buyer. He operated and owned a nearby well and sold lots of oil from his well, but according to him, did not get any oil from ours. I informed our pumper and we changed the lock to just one key. When the tank got full, he called the crook to pick up the oil, had him sign the pickup slip and we had no further problem.

I made the mistake of firing another pumper on a different well. It seemed he never went by our well. The pump was down for many days, a pickup slip was left in the box for a month, and we had not received a report that oil had been picked up. The next day, after firing him, we had an oil spill. He opened the valve and flowed 150 barrels of oil onto the ground which spread into the farmer's field. We paid the farmer $2500 in damages and cleaned up the oil and spread new topsoil. We pleaded our case to the Corporation Commission and thankfully they waived the fine.

Landowners

The property owners—farmers for the most part—are also characters themselves. They are of course interested in the drilling operation itself and some spend many hours sitting in the doghouse, talking with the crew, the mudlogger, or the geologist from the office. They want to know about the oil under the ground on their land. "Is it in big pools? Will the ground collapse when the oil is taken out? How do you find oil?" The questions go on and on. We never saw a wife visit the rig, although I am sure they were interested. Sometimes I got invited to their homes for a "good cup of coffee" though.

When I was in my first year with Pan Am, I spent probably a month overseeing, learning and studying 500' long cores cut on 10 wells. The drilling during coring is slow with lots of down time. To kill time, we were off to the little hometown café in Loyal, Oklahoma where all the locals gathered daily for gab, exchanging gossip, and to drink coffee. The combination waitress/bartender/bouncer is a young girl just out of high school. Her name is Tom Turner and she's a real tomboy—her passion in life is girls' softball, and of course she's the pitcher! I soon know all the locals and what times they typically arrive at the café. There might be a total of four houses in Loyal, Oklahoma. I have a county lease map and, from studying it, I know where everyone lives. They come into the café and Tom introduces them to me. I reply, "Oh, you live two miles east of here and one-half mile south." I don't show them my map and they continue to be amazed that I know where they live.

If you drive around the oil patch, you can see where the good wells are. The farmer has a new brick house to replace the broken-down house he inherited from his father or grandfather. Some people will not lease and don't want any holes on their land. "But don't you want a new house like your neighbor's?" I ask. They feel that a well will pollute the land and even kill the spirit of the land. Maybe they're right, who knows? Some companies are bad operators and leave lots of junk around the producing well—and especially when they plug the well, they just walk away. I tell them they can write any special clauses they like into their lease agreement to ensure their land is protected. Many wells are fenced, some with locked gates, some with planted screenings, etc. Particular areas can be protected by non-drilling clauses, or road locations can be changed. For example, an additional distance from the residence may be requested, etc. Our landman might agree to anything to get the lease. Sometimes the company can't live with a specific requirement though and a further negotiation and payment might be needed to reach a mutually-acceptable compromise.

One landowner we obtained a lease from was a very interesting fellow. He had a Slavic accent rather than the typical Okie drawl. He didn't want a well drilled on his land, but needed the money to pay a sizable vet bill for one of his dogs. He lived in an original sod house much of the year, but went to town during the coldest part of winter. There, he lived with his sister in town and busied himself with odd jobs. He also ran a few cattle on his small 40-acre ranch. He had several Russian Hounds, or some kind of breed one rarely saw. He was barefoot and wore old, dirty coveralls with a crusty felt hat. He slept on the ground each night when the weather permitted, he just curled up with his dogs. For all his rough living, he seemed fairly well-educated, and we had some good conversations. I really wanted to find some oil on his land so that he would have an income. However…alas…it was a dry hole! He was actually glad in that he really didn't want a well on his land. We did a particularly thorough job of cleaning up his drilling location, and even planted some grass seed for him before our departure from his property.

Another farmer with a half section of land got a good well. He built a new brick house just 10' from the old homestead house his father had built. Apparently sentimental though, he liked the old house and wouldn't tear it down as his wife wanted. She spent her days in the new house, and he spent most of his time in the old house—I guess he headed over to the new house for his meals! He had a rocking-chair on the old porch and she had her rocker on the porch of the new house…about 15′ away from one another.

Some farmers shot coyotes and hung them on their fences, and others hung dead rattlesnakes. Some had broken equipment scattered everywhere, and others had very clean, tidy places planted with flowers and trees. Some had fenced lawns around their houses, and some places had cattle roaming all around the buildings.

Since I came from a rural background, I was mindful of how I was raised regarding how a place should look and be maintained. In my rearing, we looked down on messy people who apparently did little in caring for their property. My philosophy is also that we don't really "own" the land, but are instead its stewards during our tenancy. When I would drive wildly down dirt roads, kickin' up plumes of dust, I would automatically slow down when passing a nicely-kept place. The County Commissioners, who often lived along some of these roads, arranged for waste oil to be sprayed on them, and sometimes they used saltwater from producing oil wells to keep down the endless dust. Today's pollution laws don't allow for these practices. We used to believe that a few parts per thousand of anything probably wouldn't hurt you. Now we argue about how many parts per billion of chemicals may be harmful! I recall being on several wells when cropduster's were spraying on both sides of the drilling rig. Hey—maybe that's why I have sinus problems!

Landmen

Landmen have a tough job. They spend many hours in County Recorder's offices determining the ownership of the mineral interests we're interested in leasing. If it's a hot play next to a good well discovery, then time is of the essence. They must contact the leaseholder and negotiate an advantageous price and terms of the lease as quickly as possible.

One of the best partners I ever had was a landman and his new wife rolled into the deal. One day, Gale Turney came into my office with a lease to sell to me. I had never met him before, but his bad reputation had proceeded him. With these stories in mind, I was cautious of him, but nevertheless instantly liked him too.

Gale and Bobbie were both in their early 60s, and both had gone through nasty divorces. Their kids wouldn't talk to them, and basically the two of them were in a low financial state at that point. But they were nevertheless two people in love and shared a buoyant, happy spirit.

I agreed to look at his lease and pulled out one of my maps covering the area near the lease location. I said, "I like the area, but it's a gas area with 640-acre spacing and your lease is only for 160 acres. Are any other leases open in the 640-acre unit?" Gale said he only had enough money to buy the one lease, but he could get the rest of the section for $25 per acre. I then replied, "I think I can sell this easily. Let's form a joint deal and we'll split the override and the lease profit 50-50." We shook hands and that was our contract for the next few years. We got over 40 wells drilled on leases that Gale and Bobbie had purchased.

From his share of the profit on our first deal, Gale bought a small RV which he took in the field with his new wife, Bobbie. Together they were also a great pair in business. Gale wrote out the lease agreements and Bobbie notarized the landowners signature. Gale talked farmer-talk and Bobbie talked woman-talk. In a hot deal with lots of competition, they wooed the farmers. Sometimes they asked to park their RV at the farmer's place since they were buying in the area and didn't want to drive back to the city. The farmer's wife would end up inviting them to stay for dinner. They would all became good friends. In the morning, the farmer or his wife would call all their neighbors, telling them what we were paying and what nice folks the Turneys were. Gale and Bobbie then just lined everyone up at the RV's door and acquired all the leases without even having to drive around to the homes of the other neighbors.

Lawyers

The oil patch requires a lot of lawyers to unravel the wild deals we put together in the field. The title of the property is very important if you're going to buy a valid lease. In oil country, minerals have been bought and sold over and over again. Sometimes in a 640-acre unit, there are thousands of owners. As referenced earlier, there may be 25–100 participants in the well in old-but-hot oil areas. Title lawyers calculate the interests of all the parties involved in five decimal places.

If the well blows out, or someone dies during the drilling, or fraud happens and more than 100% of the deal has been sold to investors, or the company didn't pay for something, a lawsuit follows. It sometimes takes years to cleanup the problem.

There are many contracts written for all phases of the drilling, completion, purchasing of oil and gas, and the distribution of the revenues. There are lawyers who work exclusively at the Corporation Commission and with other governmental bodies.

As a result of the complexity of these deals/operations, if you're in the business long enough, you will be involved in some type of lawsuit. Since I seemed to be involved often in what was categorized as a "different" kind of project, I conferred with lawyers often too. Once I took three lawyers to lunch and laid out the deal I wanted to pursue. I asked them, how can I do this deal so I don't get sued? They agreed on their advice as to how I should pursue the project and thought I was protected with this approach. I got sued anyway.

Lawyers take forever to do anything, so we would typically blast ahead without the detailed contract in hand, and sometimes we would get in trouble as a result. "You should have had this covered with an ironclad contract!" was the retort I heard after the fact many times over the years.

Jay Bond was one of our lawyers. I first met Jay in 1964 when we moved to Oklahoma. He was part of our neighborhood group of drinking buddies and I was drafted into their friendly ranks. Jay liked to play the guitar and he looks a bit like Willie Nelson, but with less hair. We had a lot of good times and drank a lot of wine together. Jay was really a farmer at heart and ran a bunch of cows on his inherited ranch; and he became our title lawyer for oil projects. We often got Jay to trade his fees for a small part of the well, and sometimes to invest too. This was slick thinking on our part. You can't receive any monies from a new well until the title opinion is done and signed by everyone. As mentioned earlier, it can often take a year for this to be accomplished, but Jay got it done rapidly as he was also waiting on revenues from his share of the well. Our investors appreciated the fast turnaround on their money and I think we secured new investments from them on other wells because of this advantage of ours.

I got beat at the Corporation Commission by a lawyer or two working for a company who I believed at the time was cheating me. I used to fantasize about them having an accident when they walked across an intersection in front of my car. (Thankfully, I have mellowed with age, and no one is "on my list" these days.)

Carroll Samara was a criminal lawyer with offices across from mine in the Cravens Building where I was located when I first went into business on my own. He was an affable, short man of Syrian descent. He had an unkempt look and always seemed to be wearing the same dark striped suit.

He went to court every day it seemed with a foot-high stack of folders and several clients. I often worked in my office until 8:00 or 9:00 p.m. Mr. Samara asked me leave my door to the hall open and, if things got too violent in his office, to yell and call the cops. Most of the offices on the floor were barely used or were populated by old, retired gentlemen—spittoons were placed outside the office doors each night for the cleaning crew to empty. The spittoons vanished slowly as the old codgers died off. I am sure that the cleaning crew was glad when they disappeared.

The arrested criminals, once out on bail (sometimes paid by Samara as a loan), filed in on 20-minute intervals. Carroll would coach them on their testimony for the next day in court. They would then typically give him the long song and dance, "I don't have any money. I got laid off or fired from work. I'm too sick to work. My government check doesn't come in until next week." It was a sad tale, but Carroll would then tell them, "No pay, no represent you. You will get two years prison if I don't get you a better deal. Give me what you have in your wallet and get me the rest before court in the morning." To which the offender would invariably reply, "I can't get the money by the morning!" To which he'd tell them in no uncertain terms, "Borrow it from somebody. You know how to get it. How did you get arrested?" They always paid up. Sometimes they complained about the deal he was cutting. "Would you rather do two to five years or just 90 days in the county jail?" They took the deal. They would each always pay him in cash before court the next day, just as he required.

One night, he comes over to my office before the next client's arrival and says to me, "Check this one out. She's a good-looker. She killed her boyfriend and I'm going to get her off!" This gal then strolls in, dressed like a high-priced hooker (which she probably was). Carroll sets about preparing her for her court appearance next week. She shows up each night for another session of instructions. "You will wear no makeup. Get rid of the purple hair. Brown is a nice color. Wear loose clothing and bind your boobs in to look smaller." He would have her practice her testimony, and practice crying and saying how sorry she was, and how her boyfriend had beaten her, and she was only protecting her life. What a performance!

I got Samara to invest $5,000 in one of our deals. He gave me the money in cash. It turned out it was a dry hole, and I was nervous he might send one of his clients over to pay us back for this dismal result. He was philosophical about the loss though and said it was his first investment in oil deals—and maybe he would just hang onto his cash in the future.

◆ CHAPTER 14 ◆ Making my own way in the biz

Once more, new partners, new deals, and a new life.

After the Davis, Northcutt and Cochrane partnership ended, I continued to have an office downtown with Bob Northcutt and Gale Turney.

A deep play was going on in the deep part of the Anadarko Basin. I had mapped the play — the Laverty Field Prospect — and liked one section for a 17,000′ $2,500,000 development redrill well. The section was home to a defunct town which originally covered 160 acres. The town had been abandoned and the abstracts that covered the town made a stack over five feet in height.

Gale (working with his wife, Bobbie) had to follow the title through heirs, and heirs, and heirs in finally arriving at the property's current owners. We subsequently spent a year putting the deal together. There was the question as to whether the county owned the abandoned roads or not, and did the original lots go to the center of the street or not? We even staked out some lots to see if we could tell. Our lawyer friend, Jay Bond, did the title work for the drilling of the well. We could not justify paying him for the hours he spent, so we gave him an override on the property along with $10,000 cash.

Jay wrote the opinion with five decimal places. Later, the operator's lawyers did not like this version and refigured the title to six decimal places resulting in $20,000 additional legal fees by their lawyers. The lawyers of the numerous participants were preparing contracts, but there were so many partners in the well, plus all the royalty owners, the well was already drilling at 14,000′ depth and the contracts were still being written. One of the partners who was having second thoughts (and had just a letter of intent, but no contract) suddenly dropped out, leaving yours truly and Gale stuck with 25% of the $2.5M well. Gale sold half of it (his interest, he claimed, but not our interest) and I was stuck with $330,000 cost of 1/8th of the well. No big surprise but that kind of ended our relationship. The well made a small gas producer and payout was several years down the road. The well has since been worked over four times in different horizons and has made money. A new deep well was drilled three years later, and of course I did not have the money for my share. So, I farmed out my interest for a small override and back-in working interest. This second well was a big one though — and my small interest brought me an initial check of $300,000. If only we had drilled there first!

After the DNC years, I initiated or facilitated the drilling of over 126 wells, all with partners — over 40 in number. In my most productive year, 1976, 20 wells were drilled. People and deals came from every direction, based on the past reputation of Davis, Northcut & Cochrane. Someone had a good lease and needed a solid geologic report to have a saleable prospect. Partners were usually only for a specific deal. Some of these individuals, like Gale and Bobbie Turney, grew into longer relationships and many subsequent deals.

I don't recall actively recruiting partners, instead they usually came to me. Some developed as partners from previous work they had done for me or DNC. A couple of landmen had deals. Sometimes I offered part of a deal for a specific service — such as the engineering for a well in a problematic area, or legal work for title searches and drilling opinions. Jay Bond became a partner in several deals for his legal title work.

I always remembered T. Boone Pickens advice: "Don't think small, sonny, but look for the big deal, the big picture." When someone came into my office with a single offset well prospect, I only wanted to take it if it could result in further drilling of more offsets. When you get 20 wells drilled in one year, several of them are offsets to wells drilled in previous years.

The only person I actually set out to acquire as a partner was Bob Marple. I hardly knew Bob and probably had only once had lunch with him when we accidently met at the Petroleum Club. Bob was a petroleum engineer and had a small company, Sooner Crude, Inc., operating only three wells he acquired prior to his position at Damson Oil Company. I had sold a couple of small deals to Natol Oil Co., which was a small but aggressive oil exploration company. Damson bought out Natol and had a different philosophy of buying production wells vs. drilling exploratory wells and only occasionally drilled offset wells. Fred (forgot his last name) was the manager and partial owner of Natol and he left Damson to form his own company. Bob Marple was promoted to office manager. Natol/Damson drilled my Tonkawa Gas discovery field (described in the next chapter) and Bob had supervised the wells completions, but I never met him during that time.

I sold a small Bartlesville Sand prospect to Fred right after he left Damson Oil. I sat the well as I usually tried to do on deals that I have sold. The well encounterd over 30' of sand with good oil shows, but appeared to have rather low porosity and permeability in samples. The well was logged and the logs indicated high water saturation with the oil, and again low porosity. Fred said, "Let's run a drill stem test before we run pipe." The DST recovered a minor amount of oil and no water, but with a low shut-in pressure. Fred said he already had the production casing loaded on the trucks, but he was going to cancel the completion as he thought that the well would never pay out. I wanted him to complete the well but he said no. "However, I will give you back the prospect leases and the hole. You complete the well or plug it at your expense. I'll circulate the well for four hours on my nickel, and then it's yours." I agreed (although I had no money to complete the well, nor even pay for circulating or plugging it).

I started down the road to Oklahoma City which is an hour away from the well. The is in an era before cell phones, so I cannot call anyone, although I could stop at a gas station and make a call or two. Most people will want to see the log and the DST before making a decision. I thought to myself, "Who can I see who can make a decision in four hours (three hours) by the time I get there?" I know many small company owners and managers who can make a quick decision. However, it's 3:00 p.m. on a Friday afternoon and most of these fellas will be on the golf course, at the bar, or headed out of town. I then thought of Bob Marple, sitting back at his desk at Damson Oil, waiting to be laid off under the new management. He'd probably like to put one over on his old boss, Fred! So I call the office and ask for him. When he answers, he says, "I was just leaving the office." I hurriedly reply, "Meet me at the Petroleum Club. I'll be there in 15 minutes and I have a deal for you which could make us both some money."

When we met up, I laid the specifics out for him and we made a 50-50 deal. Sooner Crude will operate the well, and we will jointly acquire investors and split the profits. Bob calls Fred and tells him we will take the deal. The pipe is sent to the well and by 6:00 p.m. is run into the hole. The cement trucks arrive soon after and the casing is cemented in place. Basically, we now had 30 days to pay for what we've already had done. Bob told them Sooner Crude and Cochrane Exploration would be paying for the costs incurred. Send the bills to him at Damson. Bob gives Damson two weeks' notice and we now have an oil company with a one sentence agreement.

Bob setup his own office out in the northwest part of the city, and I soon moved in with him there.

We found investors to pay for the completion and equipment. The sand in the well was perforated and cleaned up with a little acid. The well flows seven barrels of oil and dies. But we shut it in, and it flows seven barrels of oil the next day, and dies. We set a very small oil tank and wellhead with an intermitter clock that opens each day for one and one-half hours, and the well flows seven barrels of oil. It does this with little or no work on our part for the next seven years. It doesn't make a lot of money but was definitely successful. Bob and I officed together for several years, put many deals together, had a good time, but never found the "big one." [He later became a partner in acquiring the Sea Ranch Lodge and Golf Course—which is a whole different story! (See Chapter 25).]

My new business partner

Bob Marple was a petroleum engineer whose operating company, the afore-mentioned Sooner Crude, Inc., had previously been setup as an operating company—at DNC we had to have a petroleum engineer perform the technical duties. DNC only operated wells during the drilling of the wells and made deals with other companies to do the long-term operation of the producing wells. Bob was good at paying attention to the wells that he operated, and we wrung more oil and gas out of wells that larger companies would have plugged and abandoned because of uneconomic production. An engineer and a secretary can operate on a lot less overhead than the larger companies, e.g. the competition.

Once ensconced with Bob, Ann and I commuted each day from Edmond, which is basically a northern extension of Oklahoma City. The Edmond house was a new but non-descript three-bedroom tract house. We hated the traffic, so we sold the house in Edmond and purchased one on Thompson Avenue in an old integrated part of Oklahoma City. This one-story brick house with an attached two-car garage house was situated on an acre and a half lot, with many mature and stately trees. We knocked out the wall to the garage and the ceiling and combined a breezeway with the two-car garage into our bedroom-sitting room. We put sliding glass doors along the north wall which faced onto a large tree and adjacent patio area.

As mentioned, Kay had married my old chess buddy, Hal Brown. My children did not like him though, and one by one, three of them moved in with us. Brian, the youngest, stayed with Hal and Kay. I'm sure Kay couldn't give up all of her children, especially the youngest. Ann and I also acquired two foster children whose mother was killed in an auto accident while they were temporarily staying with us.

I could see this expansion coming and constructed an additional two-story wing to the house. It had four bedrooms, three upstairs and one down with a large living room. I used the large room for an office for a time.

A couple of years later, we purchased the house next door which had been wrapped up in an estate for many years and had been sitting vacant. I bought an old pickup truck and Ann, the children, and I loaded and reloaded the truck with debris from the neighboring house and yard. We made so many trips to the landfill we became good friends with all the workers there — they thought Ann and I hauled trash for a living (!).

We remodeled this other house into an office and moved Bob, our secretary Marie, Ann, and myself into our new working space. The location was close to downtown which also gave us good access to the geological library, plus numerous clients. Ann kept the business as well as our personal books. She learned drafting and we ended up doing most of our own drafting rather than rely on contract draftsmen who seemed to always be busy or couldn't fit our requests into their schedule.

Ann stayed out of the geological discussions and the deal-making Bob and I managed that as our domain. Whenever Ann became frustrated, she would quit for a while, jump on our lawn tractor and mow our two neighboring properties which comprised a total of three acres. While venting her agitation in this way, we had a huge compost heap in no time.

Since she kept the books, Ann wrote most of the checks for our business. We had three accounts, my account, her account, and Cochrane Exploration Corporation (our new company). She kept books on the corporation, split my account into business and personal, and her account into business and personal. We sometimes took working interests as CEC, Tom and Ann. As a result, she spent many days each year with our tax accountant in needing to file three separate tax returns.

Oklahoma is sometimes stiflingly hot and dry in the summer, so we wanted to have a swimming pool. Bob had one at his house which we enjoyed dipping into, so we built a 21' x 40' pool which was over 10' deep at the diving board, we didn't want the kids to hit the bottom when they dove in. We had a small pool house for the equipment and I later heard some of the kids dove off the roof of the pool house into the pool. On really hot days, I often worked in my office in my bathing suit, and took many swimming breaks. At one time, we had 10 children living with us and there were often 15–20 kids splashing around in our pool.

Occasionally I would invite a client over for lunch and drinks and, of course, to look at a deal. I made the lunches. If she liked the client, Ann would join us and get into some type of literary conversation. She had an impressive vocabulary and I often had to look up a word she was using. She carefully crafted a sentence containing two obscure words in it and then injected a four letter word for emphasis. I had one client who loved these conversations with her and would respond a couple of times a year to an invitation—one time he admitted he could only handle two of these intense conversations with her annually!

◆ CHAPTER 15 ◆   "X" marks the spot

My discovery of a gas field

Awildcat by definition is a well drilled more than one mile from production. Every oilman wants to find a new field, so wildcatting is the only way to find it. However, the odds of finding a new field are one in ten in the continental United States. During my 20-year career (1965 through 1985) I promoted many wildcats. In Oklahoma, where I spent most of my career, the odds of hitting a wildcat were one in seven. As my track record reflects, my personal odds were slightly better than the average. However, even if successful, only a small percentage of wildcat discoveries are highly commercial and lead to the development of a field.

Wildcatters never see an open area they don't automatically want to poke a hole into. They just need an excuse to do so. We geologists undertake studies to find those excuses and to then pin down the best areas to drill. We provide the invaluable service of being risk-reducers for our speculator clients. When I undertook a study, I looked at dry holes drilled out in open areas as wildcats. If they were drilled by major oil companies, you knew they had a study and some good reasons to drill where they did. A study of existing oil fields reveals, almost without exception, the discovery well (wildcat) was an edge well in the field. Companies like to drill in the center of their acreage holdings so that, if successful, they will have most of the development wells and production. Many companies buy acreage near drilling wildcat wells, hoping if the field develops in that direction, they'll have a part of the action.

My landman's brother was a geologist sitting a wildcat well for a client. My partner, Gale Turney, rushed into my office one day saying his brother had just had a free-flowing of oil on a drill stem test from the Tulley Sand. Rarely does a drill stem test produce free flowing oil, so this might be an important development. I had never heard of the Tulley Sand at that point, but I identified it as a specific sand member in a known geologic formation which contained many thin sands. I told Gale several of the area sands were highly commercial and recommended he go check out the offset acreage. He came back with the information that the company, Creslenn Oil, had purchased all the leases to the northeast and the southwest, playing the sand as a stream channel. I determined that other producing sands in the thicker geologic section were beach or bar sands running northwest-southeast.

We were able to lease two tracts offsetting to the northwest. Twenty-two wells were drilled in the field and we had interest in two of them. I invested $50,000 in one of the offsets we had purchased, and it turned out to be the best well in the field. Creslenn Oil was the operator of our well (which was about the 18th well drilled in the field). We were producing over 50 barrels of oil per day, and my investment was mostly paid back to me. (Payout is when your initial investment is returned to you. Most investors would like a three-to-one return on their investment (ROI). However lots of wells are sold with the prospect of having a two-to-one ROI, with the investor taking much of the well costs as a tax deduction.)

Pressure was rapidly declining in the field and Creslenn Oil, who operated all the wells, decided to shut the field in and begin water injection in the lowest wells. For several months the field was studied, everyone's interest was pooled together, owners had to agree, and the whole project had to be approved by the Corporation Commission. My nice 25% interest in our wonderful well dropped to .8% in the new consolidated unit. Our well became a water injection well. The field has had water injection and controlled production of the highest wells for over 35 years now and is still going, so I still receive a few dollars per month. The field has been bought and sold several times over the decades, but I've remained in the unit.

Another wildcat project

While at Pan Am in 1965, I undertook a study of the Tonkawa Formation which produces oil and/or gas from several fields in northern Oklahoma. The trend was well-developed from the western edge of Oklahoma to central Oklahoma and stretched into Kansas. Many dry holes were located between oil fields.

I think my ever-undermining boss, Howard Cotton, gave me the project to keep me out of areas the company was interested in, figuring my time would be monopolized by the project for many months.

Step 1. Of course I checked with all the other geologists who had worked the trend, got their work maps, checked our files, and read the literature. In three months' time, I had a bunch of maps and three or four small prospects.

Step 2. Howard said to check with the land department and have them see about lease availability, that took a couple more months.

Step 3. Endless review with leaders who thought the project was too small.

Step 4. I recommend the company purchase leases on three prospects. Howard says I should take it to Division Committee (he thought they'd kill the deal so he didn't want to be the one presenting the deal.)

Step 5. I take it to Division Committee and, yes, it was too small, but let's see if we can buy the acreage for a low price.

Step 6. The deal dies about six months later as the land department can't seem to purchase any leases. So, with over a year's work in the project, nothing developed and the deal was dead.

By the end of 1968, when I had left Pan American and gone into business with Bob and Herb, we decided to put together one of the prospects I had previously identified. We teamed up with a consulting landman and purchased 2000 acres of leases. We sold the prospect to NCRA, a small independent oil operator. Early in 1970, NCRA drilled a dry hole well in the center of the acreage — not at my recommended location.

By 1974, after our partnership of Davis, Northcutt & Cochrane had ended, I was teamed up with the landman I've already introduced you to several times (and my partner later), Gale Turney. Gale re-bought most of the leases (which had termed out) for half the price we'd previously paid. The following year, in 1975 — ten years since I had first mapped the prospect — we sold the project to Damson Oil. (My later partner, Bob Marple, had just gone to work for Damson. Bob was in charge of the drilling of a new wildcat well.)

The well was a discovery well for the field. I was sitting the well, as I always wanted to have some control of how my projects were evaluated and drilled. Of course it was the middle of the night as was so common in the oil fields. We had a good mudlogger on the well who I had previously contracted with on several wells. The gas detector unit didn't register any increase in gas show. The well had encountered a very thick Tonkawa sand 95' thick. The samples were mostly white sand, apparently water-bearing. I was very discouraged as the well was structurally high enough to produce gas above the water. The logger found a few cuttings with very faint fluorescence. We drilled the well 1000' deeper to look at a secondary possible pay zone. The electric logs were run and I expected another dry hole on the prospect. The logs indicated 15' of gas sand over the water. Damson sets pipe and perforates the zone. The well is completed for a Calculated Open Flow of 69 million cubic feet of gas per day rate. Yours truly is doing back flips as this is a very significant completion for a Tonkawa well in the trend. Damson then drills two more producing wells, defining the field. In total, these reserves are estimated at 12 Billion cubic feet of gas.

Time goes on—1976, 1977, 1978 and there was no hookup for the gas sales. There were no gas pipelines in the immediate area as other producing wells in the area were oil wells with little or no gas. At this point, I'm pissed and should be retired and living on just the overriding interest in the three wells. In late 1978 though, I talk with a friend, operator Al McCord, and decide to corner shoot the field from structurally upslope to the north. I invested $30,000 of my money into drilling the offset. If we were successful, we felt we could drain the field. Al said he could get a pipeline into the well. The well was a dry hole with absolutely no sand in the Tonkawa interval. We couldn't believe it! The Tonkawa sand was a beach or bar sand next to a limestone reef formation. Our well was in the limestone facies less than one-half mile from the closest producing well.

In 1979, Damson sells the field for $4.5M to Libbey Owens Glass Company. As a result of the 1973 oil embargo, in 1978 Congress passed a regulation that the industry would have to switch from using natural gas to coal, as the U.S. had lots of coal reserves. Glass companies use natural gas flames to blow glass. Libbey Owens decided to own their own gas reserves and therefore not be subject to the new regulation. I met with their contracts man when he was in Oklahoma City. He has a briefcase full of contracts, the gas is to be used in their glass production plants in Ohio and North Carolina.

The first production from these wells, which were now hooked into pipelines, occurs in 1980.

I received a check for $40 on my override. I call Damson, who is operating the wells for Libbey Owens Glass. The plan is to produce the wells at a very slow rate with a goal of producing a 50-year lifespan for the wells. I figure I'll be dead before I get rich on this one.

In 1981, Libbey Owens then sells the field for $8M. Production increases and I begin receiving $1000 per month. In 1982, the field is resold for an estimated $10M. If only they had purchased my 2% override—which, at that time, would have been worth $200,000.

The benefit$ of discovering a wildcat field

In the late 1980s, gas contracts were broken, the prices drastically reduced, and the wells themselves were producing at a low rate. My income dropped to $100 per month. Everyone seemed to make good money on the project but, me, the originator. It had taken me 20 years to get money from this project which I had originated back in 1965. Remember, in a fit of madness, I had lost $30,000 in 1978 trying to sneak into the reservoir with Al McCord? Through the years, I only received about $10,000 in total revenues—so much for getting rich on a wildcat discovery. Am I unlucky? Maybe one should never drill a gas well very far from a pipeline! But I had the joy of being correct and discovering a nice small field, surrounded by six dry holes.

◆ CHAPTER 16 ◆   Making money on oil deals

There are many avenues open for a consulting geologist to make an income from oil and gas. I have been on retainers to out-of-town small operators. They paid me a consulting fee and were provided first pick of deals which I originated. Sometimes I reviewed other people's prospects for the client. I did the geologic work on their drilling wells, especially if I had originated the prospect.

The profit on leases is one of the most important sources of income and here's how that works. Let's say I purchased 1,000 acres of leases for $25 per acre. The minimum I would consider for the leases is $50 per acre. If it's a hot area, the leases could sell for $100 or $200 per acre. Plus, I always wanted an overriding royalty interest (ORRI) on the leases. The ORRI would produce income for me as long as the well produced. I still have income on wells that initially were drilled over 40 years ago.

If the leases were burdened with excessive royalties to the landowner and others, such that the working interest partners had only 75% of the income for having expended 100% of the cost to establish the well, they didn't like to give more ORRI interest to the poor geologist (me). Then I might seek a back-in after payout, which gave me a working interest in the well after the company had paid out their investment.

If I had made a good profit on selling the leases to the company, I might take a small working interest in the well. This made the project easier to sell in that it was obvious I believed enough in its prospects to risk my own money. I tried to get carried through the drilling of the well to the casing point, thus only having to spend half of the monies needed for the producing well, but sometimes I had to bear the risk of drilling as well as the completion costs.

Working interests and royalty interests are marketable commodities. Many geologists and other oilmen sell off their interests at some stage in a well's production. A simple rule of sale or purchase is that the interests are worth three times their current annual production, giving the purchaser a return on his money in three or so years. As they're extracting a finite natural resource, wells decline in their production rates over time. However, they may be reworked in different zones, or even re-drilled within the spacing unit.

Spacing unit sizes were set by the Oklahoma Corporation Commission and related to depth and oil and/or gas production. Gas wells were spaced on larger units than oil as the rocks are more permeable to gas than oil. Gas wells of moderate to deep depth were often drilled on 640 acre spacing, which means one well can be drilled in the 640 acre unit. However, to produce all the gas and oil under the unit, increased density may be warranted and is allowed. Many areas in Western Oklahoma, originally drilled on 640 acre units, have been re-drilled on 160 acre units—thus allowing up to four wells in the section.

Since I never believed in selling off any of my interests, I have suffered through declines in prices, but also industry rebounds and subsequent increases in price. I have joined in redrills, if I liked the prospect, or I have farmed out my interest for a reduced interest. I went non-consent on a couple of operators who I thought were overcharging or for whom I thought the proposed work had little chance of success. Working Interest partners in wells sign an Operating Agreement which spells out charges, options, and legal remedies for the operator and the partners. One clause is a non-consent clause, which allows a partner to not pay for, say, a rework of a well. The operator then assumes that interest and cost. After the well has paid back the non-consent charges three times, then I (e.g. the working interest partner) was able to come back into the well. If the well workover was a poor producer, I never came back in. One time I came back in after five years.

Once the well is plugged and the initial lease has expired, the royalty interest reverts back to the landowner and all the ORRIs, working interests, back-ins, etc. terminate—then it may be time to put a new deal together and buy a new lease. Possibly the initial well was drilled in the wrong place. Maybe the well casing had collapsed and the production from the well was lost. Possibly there's a deeper horizon which has not yet been penetrated or tested.

In time, old prospects become new prospects.

Biting the bullet

Yes, there may be a value to having a contract.

The Wild West pace of the oil business kept many of us just ahead of our lawyers and their preparation of a good binding contract. We often took a deal or sold a prospect on a handshake or

Investors waiting for the electric log and deciding whether to complete the well because, at this point, they've [only] spent half the money committed to the project.

a one line agreement of understanding. The deal's here—do you want it? Can you drill it before the lease expires next week, or even tomorrow? Yes, of course we can. Sometimes we got pinched for lack of a contract when an individual or a company suddenly changed the deal since we didn't have the legal contract signed to bind it, or someone dropped out of a deal they had verbally committed to. At that point, we had to find a new partner or were stuck with that interest, or maybe the title lawyer or landman had missed an interest or not calculated the proper interest (a court action or force pooling at the Corporation Commission is needed to lock in the interest.) Time is needed to cure all these loose ends, but the lease is expiring. Sometimes the well was drilling before all the legal niceties were properly wrapped up.

Getting stuck with a working interest in a well

We used to joke about this one guy who got stuck with an interest in a well deal which was already drilling. I forget his name now, but he was the most miserable grouch I ever tried to sell a deal to. He was a geophysicist for a large oil company and (we thought) was fired or quit the company because he couldn't get along with anyone.

He put together a great deal in Garvin County, Oklahoma, and proceeded to sell interests in this deal he was going to operate himself. Because of an expiring lease, he commenced drilling while he was still trying to sell off interests in the project. A 25% investor dropped out of the drilling well—for lack of a binding contract. Mr. Grouch was then stuck with 25% of the well costs. Before the lawsuits hit, the well strikes big oil and he becomes rich and forms his own company. Grouchy but lucky apparently.

A couple of years later, he was operating his own successful oil company. I showed him a deal and he scraped it off his desk onto the floor exclaiming, "I don't take these kind of piddlin' deals!" His kid was the same age as Hadyn and pursued her around the playground. She rejected him at every turn, and we felt that we were getting back at his kid for Mr. Grouch's treatment of me. It was a good laugh.

Of course, we never thought it would happen to us—namely getting stuck with a deal. We always drafted into the bank many thousands of dollars of leases and usually were able to sell them within the 30-day bank draft period. If not, we had a line of credit at the bank which we used, along with a contract from our new purchaser, which expanded the line of credit and got the leases paid off.

I guess I was kind of cavalier about having sizable legally binding contracts. To my way of thinking though when companies or individuals go into bankruptcy, the contracts don't always amount to much—and what's there to be recovered is usually spent on the lawyers.

As a course of business, I did not sign Operating Agreements for my working interest in a well. The Operating Agreements protect the operator to a greater extent than the participants in the well. Once, I took a company to the Corporation Commission to take over a well that was virtually abandoned by an operator on the edge of bankruptcy. The company lawyer protested my right to bring the action as I was bound to the company under the Operating Agreement. My reply was; "Show me where you have my signature on the Operating Agreement." I had never signed it and took over the well and then assigned the operations to a good operator, and the well was returned to good production.

But we did get stuck for lack of proper contracts.

We had a binding purchase contract on the South Vici project for $1,000,000 for $875,000 worth of leases drafted into the bank. With our line of credit and the contract, we paid off the leases. The interest rate then skyrocketed to 22% and the client had a heart attack. (I felt one coming on too at this point!)

Then we got stuck in the Laverty deep well deal for 1/8 interest in a $2.5 million dollar well. Gale sold half of the interest as his half—not our half. Again, Gale and I only had the one line contract between us, which would have been difficult to uphold in court. That one line agreement said we would split all expenses and profits 50-50. I was not as lucky as Mr. Grouch, so I did not become rich as the result of being stuck with the well interest.

The lesson is: yes, there is value in having good binding contracts. However, they are only as good as the integrity, worth, and reputation of the participants. High risk endeavors, like the oil business, have inherent high risk…Duh!

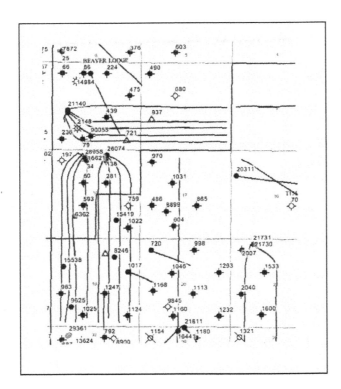

Horizontal drilling in an existing oil field
Kingfisher Co., Oklahoma

The potential problem with drilling horizontal legs in an existing oil field is intersecting the boreholes of old wells and losing the new frac (e.g. frack, short for fractured treatment) into them. If the well is producing, this will kill the well, or channel water where you don't want it.

It appears that most, if not all, of these wells have been plugged. Modern plugging requirements include ensuring cement covers all the old perforations, so there should not be a problem. If instead a well was plugged many years ago, then these perforations may not be covered with cement.

◆ CHAPTER 17 ◆   Oil & gas deals can be ri$ky

I guess almost all of us are gamblers at heart!

Nearly every oil well is probably drilled with several investors. Big companies join with other firms to spread the risk or share the cost. Sometimes small companies or private individuals own leases or minerals in the drilling unit.

First, the drilling unit is formed. The size depends upon the depth of the producing horizon, and if gas or oil is expected. Shallow wells can be drilled on very small spacing sizes. For moderate depths (5000' – 15,000') oil spacing is 80–160 acres, and gas spacing is on 640 acres. The new horizontal wells in Oklahoma are drilling a single well on two section spacing (1280 acres). Let's say you own 10 acres of mineral rights in a 640-acre section spaced for gas. You can lease your interest for so many dollars per acre and retain a $1/8^{th}$ or $3/16^{th}$ royalty interest in the well. You will then receive $1/8^{th}$ times $1/64^{th}$ part of the production, decimal interest calculates .001953. Or, you could join in the drilling of the well by paying $1/64^{th}$ of the cost of the well (10/640 acres) — for which you will then receive $1/64^{th}$ of the production (.015625).

Say the well costs $1,000,000 to drill and complete. It will cost your interest $1/64^{th}$ or $15,625 to join in the well. If the well sells $50,000 per month worth of oil & gas, it will take nearly two years to pay out the cost. You will receive $781.00 per month on your share of the well. If you had leased your 10 acres to the drilling company, you would receive $98.^{00}/month, but at no cost to you. If you had leased the ten acres for $500.^{00} per acre, you would have received $5000 up front.

The above scenario is the kind of return on investment that many of us receive from wells. However, every now and again the BIG one comes in. One of my best producing wells brought me $1000/month for 10 years (e.g. $120,000) from a 4 acre interest I owned in the section.

In oil boom times, everyone wants to be an investor. In poor times, few wells are drilled and acreage lease prices are cheap. One small company I sold deals to formed an oil company because they owned thousands of acres of mineral rights in central Oklahoma. During the beginning of the Depression in 1929, the company owner's grandfather, named Borelli, left New Orleans and moved to Oklahoma. He had a large stash of money (we're not sure why). All the farmers were barely hanging on and needed cash badly. Borelli bought half their mineral rights for a very low price. There was no oil production anywhere near that area at the time. Oklahoma City field had been discovered with a billion barrels of oil reserves. Everyone bought and sold leases in and around the field. So Borelli bought all these mineral rights 20, 30, 40 miles west of the big play — the locals must have thought he was crazy!

It did take 30–40 years before oil drilling came to that area where he owned those mineral rights. As a result of his astute foresight though, his grandchildren became very wealthy and started their own oil company, Brown & Borelli, Inc. with the income from those mineral rights. At first, they just sold leases on their mineral rights. As their income grew from oil production on these longtime mineral leases, they decided to form an oil company and take working interests in drilling deals involving their minerals. This later progressed into drilling and operating wells, and acquiring an investment group to spread their risk. Initially they just drilled offsets to good wells drilled adjacent to their mineral holdings. I became involved with them, as they became partners in deals I promoted. Soon I was showing them oil deals and they were branching out into other areas outside of the locales where they owned mineral rights. I actually sold them a couple of wildcat prospects. One was moderately successful and resulted in drilling six or eight offset wells.

Other inve$tors

In my oil days, the investors were often the children of people who had made money in the oil boom of the 1920s and 1930s. Doctors and lawyers were investors too. I got my lawyer friend, who did the title work on our oil deals, to invest. A side benefit was we got our title work done in record time as Jay wanted to start receiving income from the new well. In contrast, I've been in wells that took over a year after the well was finished for the legal work to be completed.

We had an investor who had the franchise for all the school buses sold in Oklahoma. We also had an investor who owned a Jack-in-the-Box fast food restaurant. He had no money during the boom times, but invested with us when the industry was having a tough time. I asked him about it and he claimed during the tough times all the oil men opted for fast food at lunchtime so he had cleaned up whereas during boom times they all lunched at the Petroleum Club!

Often investors want a tour of the drilling rig so they can see how their money is being spent.

The operating company sends out daily drilling reports to all the working interest partners so that they're kept informed as to when we will reach total depth (TD) and are logging the well. If it's close to town, they'd come out to the well. The electric logs are run in the well and printed in the logging truck. The logging engineer calculates all the potential pay zones. The mudlogger has his log of the samples and the gas and oil shows encountered during the drilling. The geologist (myself, if it's my deal) will be on location to see if his interpretation is correct (e.g. the reason for drilling the well in that location). The company owner (if it's a small company) or representative will also be there. Then everyone puts their heads together to see if we want to test a specific zone or to run pipe when it appears most likely it'll be a productive well. At this point, only half of the total investment has been spent, so if it's a definite dry hole, then it's time to plug the well and save half the money and put it toward the next deal instead.

By this time, it's usually midnight or later because nothing like this happens during daytime office hours. At least in my day, electric logs had the propensity of not being completely definitive. They would indicate oil, but usually measured water content too. All sand grains seem to have a coating of water and the oil is found in the porous spaces between the grains—thus most zones calculate as having some percentage of water. A zone may produce water free with salt water saturations of over 50%. Or they may produce oil and water, or just water. And so the agony begins. I want to see pipe run on the potentially productive zone because we have more acreage in the prospect and, even if this particular well is a marginal producer, it may lead to better production. The investor—if he agrees to the completion—has to put up more money for the additional effort to be undertaken. Many investors use oil drilling deals as tax write-offs, so a dry hole, although unfortunate, may actually provide a tax benefit to him.

One deal I had sold to Brown & Borreli, Inc. had three investors out at the rig helping us make the decision. We finished logging around midnight, and two hours later everyone is still trying to make up their minds about whether to set pipe. I was tired and also felt sick after being on location for three days looking at the samples, picking the total depth of the well, and supervising the electric logging. I went outside for fresh air and threw up. I went back in and declared that if they want to walk away, I will take over their interests and attempt completion of the well. That made their minds up and everyone said let's go ahead and run pipe. It turned out to be a nice well and produced minimal water with the oil.

Another investor was Dr. Freede who owned small pieces of mineral rights in several counties, so he kept showing up in our well deals. Each time I would approach him and ask him to join or sell us a lease, his reply would be, "Take me to the Corporation Commission and I will make up my mind."

So that's what we did and he usually joined in the well. We completed several nice wells and it was always the same answer with him. One time, I told him, "Everyone else has made up their mind and we don't need to go to the Corporation Commission, can you just make your decision now because we have some expiring leases and need to drill this well soon." He went ahead and joined and we made a nice well. On the next deal, I called and asked him, "Do you want to join in this one? In fact, we have some more interest to sell in the unit. Do you want to take a bigger slice of the pie?" He said "Yes" straightaway and became a member of our investor group for that deal too. Dr. Freede was not a geologist, but he had an uncanny ability to estimate how strong a producer a given well might be, or how risky the well location was. Once I even moved the well location slightly to a spot he suggested, and that turned out to be a good well.

Legally forcing my way into a well

Keith Walker had a small company and was drilling shallow oil wells near a field that I had a small interest in. He was drilling for oil on 10-acre spacing and hit gas instead of oil. I had a lease next to his well in part of the 40-acre track surrounding his well. Gas at that depth was supposed to be spaced on 40 acres, but was not spaced in that area. I approached Keith and said, "Hey, Keith, you're draining my gas from under my lease. It should be spaced on 40 acres instead of 10." He replied, "It's just a little well, don't worry about it. It probably doesn't extend under your lease." I then filed with the Oklahoma Corporation Commission to space the land on 40 acres for gas, and I also filed a force pooling to put his well in the 40 acres and put me under the producing well. The Corporation Commission determines the optimum size of the spacing unit and all owners within that unit have to make a decision to join, farmout, or sell a lease to the named operator of the well. All the royalty owners under the spaced unit receive royalty payments. Thus, I was not only wanting to be included in the well production, but the action also produced royalty payments to 30 more acres than under the original 10 acre spacing.

At the Corporation Commission proceeding on the matter, I called Keith Walker to the stand as my only witness. He seemed a little confused by that and definitely wasn't prepared to testify — especially as my witness. My lawyer asked him about the well, "Is it, Mr. Walker, a gas well or an oil well? Does the well produce any oil?"

"Well, I guess technically it's a gas well since it doesn't produce any oil. My offset wells from the same formation are all oil wells."

"Mr. Walker, did you have a geologist's report and map showing the sand trend?"

"Well, yes, after the well was completed, I had Bob Stroud prepare a new map."

"Mr. Walker, do you have that report here with you? "

"No, I don't."

"Do you recall how Mr. Stroud mapped the sand trend?"

"Well, yes. I think he showed the trend going northwest and not going under Mr. Cochrane's tract of land."

"Mr. Walker, do you remember Mr. Cochrane asking you for a copy of that report?"

"Maybe he asked me right after the well was drilled."

"Did you tell him you would have Bob Stroud send him a copy of the logs and the report?"

"Well yes, I guess I did."

"Mr. Walker, is this a copy of the report and map?" My lawyer hands him the report and map which shows the sand trending under my lease. The Corporation Commission, without further testimony spaced me into the well. Since the well had produced for the year it took to get it to the Commission and the pressure was down, it was ruled that I would pay my portion of the completion cost, but none of the drilling cost. The completion cost was approximately one-half of the total well cost to drill and complete the well. I got a good deal and Walker got a good deal in that the initial production had really paid for most of the total well costs. The well produced gas for three or four years after I came in, and I made about a two-to-one return on my investment.

Once a person has invested in an oil deal, it gets into their blood. On top of that, their names show up on drilling opinions on wells that we might have an interest in. And then companies sell lists of their investors. I cannot believe that over 20 years later, I still get phone calls nearly every day — certainly every week — from someone with a hot oil deal in Texas, Kentucky, North Dakota, or even offshore. I also get pitched on stock deals, a movie deal, or even a clean energy deal. I received one the other day for a new refinery to be built in North Dakota (the first privately offered refinery to be built in the U.S. This might be an alternative to the pipeline that's receiving so much flack).

One of the functions of the Oklahoma Corporation Commission is to legally determine who owns the minerals, and even to force mineral owners to choose an option when a well is proposed by the major lease holders in the unit. The choices for a mineral or lease holder when "force pooled' are to: (1) lease their interest in the drilling unit to the operator (price and retained interest are determined in the action), (2) farmout their interest in the drilling unit to the operator, normally delivering a 75% lease (percentage may vary), or (3) join in the drilling of the well for their mineral or lease percentage in the spaced drilling unit.

In many areas in Oklahoma the mineral interests have been bought and sold multiple times. They may have been split out to numerous heirs of the original landowners, and further down the line to others. Often a drilling unit will have several minor interests in which the owners addresses or whereabouts are unknown. The operator is assigned these interests, but the revenues from production are held by the state. These monies can be acquired by filing a form with the agency's Unclaimed Monies Department. Although my personal mailing address has not changed for many years, I have found that companies may lose one's address, especially as companies buy and sell their myriad interests in mergers, buyouts, and bankruptcies. So anyone can check online to see if their name appears on their list of unclaimed monies from these transactions.

♦ CHAPTER 18 ♦   Wild days in the oil patch

It's always been boom or bust!

Throughout the long history of the oil industry, the prices of oil and gas have fluctuated wildly. In the old days, it was mostly a function of the price and availability of crude oil. In later years, political factors affected pricing, as well as reserves. Refining capacity or changing chemical additives during the winter and summer seasons, hurricanes in the Gulf of Mexico, wars in the Middle East, ruptured pipelines, etc. have all affected price and availability of these key commodities. In 1973, OPEC (Organization of the Petroleum Exporting Countries) decided to takeover and control the crude oil pricing.

(Another one I just missed — I was in the process of purchasing an old oil field I thought could produce a lot more oil with a proper study and a secondary recovery project. When OPEC changed the price, the field was no longer available, and our signed deal was negated.)

As mentioned earlier, in the late 1920s, Oklahoma City field was discovered with a billion barrels of oil reserves. Every railroad tank car in the U.S. was taken to Oklahoma City and filled with oil. Huge pits were dug over the landscape and millions of barrels of oil filled these pits. The price of oil dropped to 10¢ per barrel, which meant it cost more to drill for the oil than it was actually worth on the market. The Oklahoma Governor called out the National Guard and closed down the oil field and all drilling activity saying, "We are not going to sell our oil for 10¢/barrel. We are going to regulate these wells in daily production, not have open pits of oil, and cleanup the oil field."

This was the beginning of the regulations that now control the petroleum industry. However, this was done at the state level rather than the national level, which is a problem to this day — with different regulatory rules in play in different states. The big oil play in North Dakota — again with few rules — is just now having to develop policies to govern the industry there.

Fracking was begun in the 1950s and, if it had been thoroughly studied, could have developed rules to safeguard the environment in decreasing water contamination and preventing earthquakes. Of the many wells I have studied and drilled that were fracked, the pollution of the fresh water table has been zero. Frack water follows the path of least resistance and can funnel along a natural crack or fault. One well I had an interest in, we gave the well the standard frack (at the time) of 10,000 barrels of water treated with sand to prop open the created fractures. We never did produce any oil from the well. The well flowed back the 10,000 barrels of our frack water and another 10,000 barrels of someone else's frack water. Obviously, a fault or large crack was funneling the water from a long distance away. Since the frack was nearly two miles below ground, the fault did not extend to the surface of the earth. However, it is possible for shallow wells to be fracked into the fresh water table, or possibly to the surface. Another danger is fracking into old well bores of improperly plugged wells, or wells with failing or rusted casing.

When I was hired by Pan American Petroleum in 1965, prices were low and the industry was cutting jobs. OPEC spiked the price globally in 1973 and domestic oil looked much better in the late 70s and 80s in light of these now-higher prices. Small deals were drillable as a result, and small outfits sprang up everywhere. Drilling funds were raised in the stock market. Promoters showed up daily at our offices looking for deals. It was common to meet a movie star, a football hero or coach at our local bar — each of whom were being promoted or doing the promoting for a large independent oil company and their deal(s). Bartenders, car dealers, and a host of folks in unrelated jobs were suddenly oilmen looking for deals and raising money. I sold a three-well project to a furniture dealer. These were successful and 30 years later, one well is even still producing. Basically these types of operators had no geologists or petroleum engineers. They were simply money- raisers and contracted out the drilling and operations of the wells themselves. Some of them had fancy offices and operated wells with limited office staff. However, their lack of industry expertise in evaluating oil deals led to a high percentage of dry holes and marginal wells resulting from their deals. The smarter ones sold off their production and closed their offices when the boom period began ending. However, others went into bankruptcy on the heels of their high stakes gamble.

During these boom times, natural gas prices in some areas still remained rather low, primarily due to lack of pipelines and/or markets. Shallow oil projects were basically used up. Deeper gas prospects became the target in the 1980s. Natural formation pressures increase with depth due to the overlying weight of the bedrock. Rock formations are 50 times more permeable to gas than to oil. The problem is when wells are accordingly drilled to deeper depths, the cost rises exponentially. In Oklahoma 20,000' deep wells were becoming increasingly common and some wells even reached or approached 30,000' in depth.

In the late 1970s, Robert A. Hefner III lobbied Congress to get special pricing for deep gas. Even today, gas is cheap compared to oil — currently at $1.30 to $1.93/mcfg. Hefner got the deep gas prices up to $8 to $10/mcfg, and the deep drilling exploded thereafter as a result of his help in getting the Natural Gas Act of 1978 passed.

One of my friends, Clayton Lee, began constructing deep gas drilling rigs at a cost of $6M per rig. He soon had six of them under construction. Every deep drilling rig in the state had contracts for more than a year. Lease costs escalated to thousands of dollars per acre for short term leases. The banks decided to finance drilling rigs as less risky than loaning for lease costs or drilling costs. We all promoted each other and ourselves.

I authorized a partner, Ron Eddington to handle the leasing on a deep project and we had $875,000 worth of leases drafted into the bank (this was during the Jimmy Carter years). Due to the dicey economy at the time, the interest rate soared to 22% and shortly thereafter, we were in trouble. I got my friend with the huge loan for his six rigs to become a partner in the deal, and he agreed to purchase the leases for $1M. He also agreed to pay the bank interest until the money became available (in three or four months?). Then things went downhill.

Money was due at all the banks — Seattle First, Continental Illinois, and Citi Bank held most of the notes on the drilling rigs and were now in trouble…because the boom was ending and the rigs were out of work.

It had been a wild time for a while. A banker friend was photographed drinking champagne out of a call girl's high-heeled shoe. Cowboy bars were the liveliest places to bring out-of-staters who were visiting. The bars had these bucking bronco machines and girls wearing cowboy boots and hats would ride them and sucker guys to take a ride. Of course, the machines had several levels of bucking. You started slow and they increased the speed until you were bucked off. People broke arms and were severely bruised — a great laugh from the drunken audience applauding!

Penn Square Bank in Oklahoma City was shut down and people were fired and also prosecuted for their shady deals. I banked at the First National Bank in Oklahoma City for many years and had borrowed and repaid a lot of money to them over the years. Most of their loans were for real estate. All the new oilmen were suddenly building 10,000 square foot houses. When the boom busted, the conservative First National Bank suddenly had a lot of big homes in foreclosure. My bank loan officer was Greg Jones and he had actually helped us with the million dollar loan. He was very conservative and had always way-undervalued my assets. He wanted statements from all of the partners in the project before he would make the loan. Suddenly one day he was gone, and I had a new loan officer assigned to me named Trudy. I made up a little ditty about Trudy that I sang every time I left her office. It began "Life is too short to have Trudy, to me she is very rude-y ..."

I then looked for a new bank, especially after hearing Greg had moved over to Penn Square Bank. I stopped by his new office and told him of my troubles with Trudy. I asked, "So shall I come over here to Penn Square?" Greg said, "No, don't come here, you don't want to be here next week. Go see my friend at Republic Bank." The next week the banking regulators shut down Penn Square Bank! Greg had been hired by the bank and the regulators to try to save the institution and he continued to work there for more than two years to attempt to make some sense of the boom days. It seemed that the money all came out of the three afore-mentioned big banks. They suffered great losses and several bankers lost their jobs. A couple were also hauled into court for their dubious loaning practices.

Thankfully, Ron and I escaped the bust and sold our million dollars of leases to a Texas oil promoter, a small Oklahoma Company, and an Arab banker from Cypress.

The Hefner family starting with Judge Robert A. Hefner and extending to, his son, Robert A. Hefner Jr., and Robert A. Hefner III, have a long history in Oklahoma and Texas in the oil business, oil and gas law, politics, and the arts.

Judge Hefner was one of the first to establish oil and gas laws in Texas in 1903 after the discovery of the Spindletop giant oil field. He moved to Oklahoma where he developed a successful oil company. He was named Justice of the Supreme Court of Oklahoma and served from 1926 to 1933, and wrote some of the Oklahoma oil and gas law. From 1939 to 1947, he served as mayor of Oklahoma City. Schools, roads, and Lake Hefner reservoir are named after him.

Robert Jr. also practiced oil and gas law and further developed the Hefner Oil Company into a very successful enterprise.

Robert A. Hefner III continued in the family business after receiving a degree in geology from the University of Oklahoma in 1957. He founded the GHK Company and pioneered ultra-deep gas exploration and development. He drilled well record depth exceeding 31,000' depth, a record which stood for over 30 years, but was finally bested in 2004. In 2007, GHK was the first mover in Oklahoma to drill horizontal wells in the Tonkawa Formation.

I ran into him from time to time at various industry gatherings; after all, we were a relatively small community of players all told. In addition to leading the way in ultra-deep drilling, he was also the first to engage in horizontal drilling, and seemed to me to clearly be a man of vision overall.

♦ CHAPTER 19 ♦   Oil spills n' blowouts

Disasters in the oil patch are primarily caused by human error. Mother Nature will get you if you don't stay on alert—the deeper you drill into the earth, the greater the pressure you must control. Hydrostatic pressure is the weight of a column of water and increases at the rate of .45 psi per foot. At 10,000' depth, the column of plain water exhibits 4500 psi. However, at the same 10,000' depth, the formation pressures may be much higher. To increase the weight of the mud column to hold the pressure, additives, such as barite, are added to the mud. In the old days, wells were drilled with water and they simply blew out when they hit high pressure zones. The result was a sizable spill of oil, and often, an explosion with the rig burned up—in which, hopefully, no one was killed.

Early in my career I watched a drill stem test of a high-pressure gas sand. In a drill stem test, a packer is set just above the zone you want to test. The mud pumps that pump heavy mud into the drill pipe are shut off. When a valve is opened, the pressure of the reservoir is released into the drill string of pipe. When you are drilling, the mud is pumped down the pipe and the pump pressure and reservoir pressure push the mud up the hole and outside of the drill pipe. For the drill stem test, the roughnecks attach a pipe to the drill pipe and aim it into the reserve mud pit.

In this first test I watched, the crew had tied the pipe with a rope to a railing on the drill floor. The pressure was maybe 8000 pounds per square inch (p.s.i.)—this immediately broke the rope tying the pipe down and it began to violently flop around. Everyone ran for their lives! The driller closed the valve and the pressure was contained. If he'd not done this immediately, the gas would have hit the hot engine exhaust pipes and we would have had an explosion resulting in the rig undoubtedly being destroyed!

Undeterred, we wanted to continue with a new test. This time the pipe was chained down to the drilling rig versus a flimsy railing. The pipe was choked down to a 1/8th" choke (hole) to reduce the amount of gas produced. The valve was opened and the well then screamed and shook with the high-pressure gas flow. We obtained our test data and I learned a lesson that I always insisted on whenever I ran a drill stem test—namely, chain down the drillstem test flow pipe and be able to close down the test at a moment's notice.

Wells drilled in high-pressure gas areas usually set a string of pipe above the high-pressure zone. The pipe is cemented into place, and the cement and casing protect all the uphole zones from the high-pressure zones below. Several blowouts occurred during my oil days in Oklahoma resulting from the pipe not being set at a proper depth above the pressure zone. Out-of-state operators may not be aware of the severity of the high-pressure problem. A couple of operators who had blowouts were trying to save money by not setting the intermediate casing—they thought they could contain the pressure by mud weight alone. The problem was that the heavy mud flowed into lower pressure zones and the gas came into the hole, which lightened the mud akin to removing a champagne cork— and boom, a blowout!

One such well, near Canton, Oklahoma blew up and burned the rig (see the book's cover photo which was taken three months after the well first exploded). We could see the dark and ominous smoke billowing into the sky 125 miles away while sitting at the Petroleum Club bar. Once again, the dumb operator had cut corners and not surveyed the hole as he drilled the well. It burned for an entire year—billions of cubic feet of gas. Relief wells were drilled to intersect the wellbore, to pump mud in and kill the well. But we didn't know where the wellbore was precisely located. After the year had passed with this situation continuing—including the drilling of the third well—the wellbore was finally found and the renegade well was killed at last.

For one well I had an interest in, I take most of the responsibility for losing it. The

The drilling crew. The guy in the cap is the pusher or maybe the driller.

pusher and I had gone into town to eat dinner as the well was drilling slowly and, given our pressure zone, we thought it would not be reached for many hours yet. The driller was instructed to raise the mud weight to 10.6 pounds per gallon before we reached the zone. However, the well drilled much faster than anticipated, and the mud weight was not brought up to our recommended weight. The well drilled a 12′ gas zone—we were expecting maybe 6′ of pay. The well tried to blowout and the crew shut in the blowout preventers. This effectively cut most of the gas flow. The surrounding rock formation blew into the hole and stuck the drillpipe so that it could not be moved in any direction. They worked on it for days and ended up leaving the drillpipe in the ground. Luckily, the rig was not lost, but the cost of this unexpected situation was certainly high with the loss of the drill string of pipe as well as the time expended. A new well was drilled offsetting the lost hole. It turned out to be one of the best wells in the field—and I lost a good client.

Poor equipment is another cause of rig failure and loss of a hole or a blowout. Drilling crews don't like to test the blowout preventers, which may often need adjusting, or a replacement part or something. I always required the drilling be shut down before we reached a possible high pressure pay zone and the blowout preventers be tested. Several times in my career I have been on a well, a zone has tried to blow, and the blowout preventers closed. The mud was then brought up in weight to handle the problem, and drilling continued.

I sold a prospect to a small operator named Phil Boyle. He "poor-boy'd" everything and got a rig to drill the well which was old, dirty, and underpowered. I paid for a 1/32 working interest in the well to get him to take the deal. Later, he calls me and says he has cut 50′ of Layton sand, which was a good producer in the area and situated several hundred feet above our target zone. He decided to drillstem test the interval to see if a water contact was in the bottom of the sand. The result was a great test flowing free oil with good pressures. We were ecstatic and literally jumping up and down with joy on the rig floor. The drilling crew then proceeded to drop the drill string in the hole as their brake shoes were worn out, but they could not fish the drillstring out of the hole. Boyle fired them and told them to get their rig off his location. He then got a better rig from a different drilling company and they then whipstocked (side-tracked) the hole to get around the stuck pipe in the hole. The new bottomhole was 150′ east of the old hole. The Layton sand was only 30′ thick with lower porosity. It made a fair well at a high cost to everyone, but I believe the original spot would have made a great well!

Failed pipelines are disasters so those are the ones that make it into newspaper headlines. In reality, pipelines are everywhere, especially in oil country. Every producing gas well has a pipeline to the well. There are 2.5 million miles of pipelines everywhere throughout the U.S. When these lines are first put into the ground, they are tested for high pressure, and present no problem. Over the years however, wells are then bought and sold. New operators acquire old equipment on stripper type wells. The main lines leading to pumping stations, refineries, and towns are supposed to be checked often—but the truth is they may not be. Some oil and gas lines have been in the ground for more than 50 years. The oil inside the pipe probably prevents rusting, but what about the outside of the pipe's casing? Gas pressures decrease with age in producing wells. During the later stages of a well's life, a compressor is put on the well to buck the line pressure in the pipeline, which is higher than the natural pressure in a well. It is not economical to replace a gas or oil pipeline on an old well which barely produces any revenue. There are lots of minor leaks everywhere, but with low pressures and low production, the damage is typically minor. (You of course will not believe it's "minor" if it's located in your backyard! The company will fix the line and pay you a few dollars for the damage, and then it's back to business as usual from their standpoint.)

Offshore pipeline failures can be more of a problem in that the oil and gas are scattered everywhere in the local ocean's environment. The pressures in play may be high, which also adds to the size of the spill. Another factor: sea water is also more corrosive than the corrosion a pipe is exposed to when buried in the ground.

The major oil companies find and drill the flush initial production. Pressure and production declines as the oil and gas fields are drained of their reserves. The majors then sell the wells, pipelines, and refineries to independents who have less economic backing to cover spills, not to mention more limited resources in terms of time and money needed for testing in order to prevent spills. As shared, I have interests in wells which have been producing for more than 40 years. A new company buys the old well, sprays some paint on the rusty equipment and the equipment itself continues to rust underneath the paint. They tell the landowner, they have cleaned up the well. They remove old, rusty, abandoned parts and tanks so the farmer is now happy—but perhaps he's been deceived into a false sense of security.

The U.S. has 2.5 million miles of oil and gas pipelines. The states with the most pipelines are Texas, Louisiana, and California. Half are over 50 years old. Only a percentage of the lines are inspected or regulated by the federal government. The larger, high pressure transmission pipelines are supposed to be the most rigorously inspected. The San Bruno, California explosion in 2010 was the result of a defective seam weld in a pipeline constructed in 1956, before current regulations. A high pressure test or a robot 'pig' might have detected the seam corrosion and prevented the accident. The United States Geological Survey registered the explosion and resulting shock wave as a magnitude 1.1 earthquake. Eyewitnesses reported the initial blast "had a wall of fire more than 1,000' high," and there were eight fatalities.

Most problems are caused by corrosion, weld failure, or result from impacts to a pipeline by construction equipment.

The least-inspected lines are gathering lines to wells. The pressures are higher in new wells, but the gathering lines are usually new and can handle the pressure. Urban and suburban homes obtain natural gas via distribution lines. Some of these are very old, but the distribution gas pressures are low. Leaks are very common though, often occurring within the home, or are caused by construction equipment in residential areas.

The Trans-Alaska Pipeline (e.g. Alyeska Pipeline) has been operating for 40 years since its first production in 1977. It is 800 miles long and stretches from Prudhoe Bay to Valdez. It had to overcome more objections and required more environmental studies than any such project built at that time. It was authorized by federal legislation, and was bolstered by fears of losing oil from OPEC following the 1973 embargo. There have been some minor spills over the decades, but it is a highly regulated and inspected pipeline. The oil revenues it's produced to date are unparalleled by any other such pipelines from Alaska.

The proposed and highly controversial Keystone Pipeline is to be constructed from Canada to Texas refineries and is receiving a host of legal and environmental challenges. The Dakota Access Pipeline (Bakken Pipeline) runs from North Dakota 1172 miles to Southern Illinois. It too received much public objection, especially from the Native Americans in the region. President Trump approves of these projects and, as of the time of this writing, a lawsuit from a key tribe is still pending in federal court to determine whether its operation can be halted pending an environmental review by the U.S. Army Corps of Engineers.

◆ CHAPTER 20 ◆   Hitting the target—defining a stratigraphic trap

Understanding sedimentary features on the land surface helps us find underground features in the subsurface.

Imagine you are walking along a sand beach on the California coastline or some other coastal region in the distant past—although you probably aren't actually a human being in this scenario. In your three-mile hike of the day, you walk along a sand beach, say 400'–500' wide. The beach changes in width, winds into coves, and curves around rocky points. The sand underfoot varies from a coarse sand, to a fine sand, to a section of windblown sand with small dunes. The beach disappears at the rocky points of land and occasionally becomes a beach composed of rounded pebbles, and even boulders.

Five million years later, as a descendent of the previous walker, you return to the area that was this beach all those eons ago. The beach is now covered with two miles of rock strata, but it's important to find because it contains oil. You determine the beach is down there covered with the two miles of younger rock. You surmise this from examining the data acquired in a dry hole drilled by the ABC Oil Company. Their well encountered five feet of tight beach sand with oil shows overlying 15' of white saltwater-bearing sand.

You say, yes, I remember this sand beach, it used to be 500' wide, ran basically north-south, and meandered in and out of coves around rocky points. The points separate the oil contained in the beach sand into different reservoirs with, probably, different oil-water contacts. These are stratigraphic traps and are important oil and gas producers around the world.

So, back on the surface, how does one drill two miles into the ground and hit a narrow 500' wide beach and find oil above the water contact? It often takes more than one well to find the oil. If we drill downdip to the west from the ABC Oil Co. well, we will have saltwater. If we drill updip to the east, we may miss the beach sand completely. Along trend we want to find the beach updip from the water, if there is actually even oil in the sand. Luck comes into play here too in conjunction with good mapping and postulating (imagining) how the original beach ran in ferreting out where the oil is. This is what we call a stratigraphic trap (focusing on the study of rock layers e.g. strata) as opposed to a structural trap where oil is found only at the top of the structure (anticline).

Structural Traps and the Role of Seismic Studies

The folding of rock strata commonly produces anticlines and synclines. Whatever oil and gas is trapped in the rocks migrates updip (upslope) to the tops of the anticlines. Saltwater, being heavier in weight than oil, occupies the voids between the grains and is found in the lower part of the structure. Geologists review the records of dry holes for shows of oil and gas tested along with saltwater. Updip from these wet wells, which also had shows of oil and gas, indicate where we should look for the elusive oil field. Anticlines are often the prime target of the search.

In the early "oil patch" days, large anticlinal structures were sought and found via mapping surface geology. One such discovery was the Oklahoma City Field which covered a township-and-a-half on the east side of Oklahoma City. Houses now cover the field. The Oklahoma Capitol building had angle-drilled wells underneath it. All the wells in the immediate vicinity produced over one million barrels of oil each in the 30 or more years of their production. One well may still be there, kept for historic reasons. In the 1970s, I was present when the Oklahoma City Geological Society installed a large enameled metal sign showing the geologic structure under the Capitol. The large anticlinal structure was formed during the Pennsylvanian Period as a large structural uplift on the south end of the buried Nemaha Ridge. Over 7500' of pre-existing formations were eroded off the rising structure. Rising sea levels (really rising and falling sea levels—known as cyclothems) onlapped the structure with 2500' of thinning and non-deposition of lower Pennsylvanian formations. Finally, the Oswego Limestone of Des Moinesian Age covered the Cambrian Age Arbuckle Formation. This produced essentially a structure with two miles of vertical difference from the adjacent rocks in the Anadarko Basin. Oil and gas are produced over and on the flanks of this structure from several different formations.

Rarely do we find a structure in the Mid-Continent of this magnitude. (NOTE: California terrain is much different with large faults and structures everywhere. I recently authored a book designed as a primer for the layperson of the geology along an 85-mile section of the coast there: *Shaping the Sonoma-Mendocino Coast—Exploring the Coastal Geology of Northern California*, www.RiverBeachPress.com.) A small wrinkle in the rocks in the Mid-Continent with less than 100' of closure can produce significant oil and gas reserves. Geologists draw maps from existing well data looking for potential anticlines. A change in the structural gradient, e.g. an anticlinal nose on our map, may indicate that a closed anticline—and an oil field—is nearby. Many wells are drilled on this information alone, without conclusive evidence that a closed structure is actually present. Sometimes, two or three wells later, the field is found. Another way to delineate the structure is to shoot seismic lines. Earthquakes produce seismic waves which have been used to determine the various layers of our Earth, including its core, mantle, and crust boundaries, etc.

Seismic waves are also produced from volcanic eruptions and from large explosions, including nuclear tests.

Geophysicists use seismic information in making maps to pin down the tops of structural features and increase the accuracy of the target well. Seismic technology continues to change and improve over the years. However, the basic principle is to send an electro-magnetic or sound wave down into the ground and then examine the reflections which bounce back to a receiver located on the surface.

During my days in the oil field, there were two methods in use to acquire seismic records. A line or several lines of geophones were placed on the surface, or in small holes, to receive these seismic reflections. Waves were generated from exploding sticks of dynamite placed at different spots in or adjacent to the geophone lines. The second method was to use "thumper trucks" in which a large weight or hammer hit the ground to produce the wave as the truck moved over the surface, dispensing repeated thumps.

Electronic equipment recorded the data reflecting these waves. Seismologists in the office then prepared seismic maps from the interpretation of the data. However, several problems existed which complicated these seismic interpretations. We wanted a seismic map to be drawn on the target formation, which may or may not be a good reflecting formation. Determining which reflection belongs to which formation may be accomplished using different filters or different wavelengths, or other manipulations. (Ah, the secrets of the geophysicists—it seemed whenever I asked to see a specific line, they always brought out a different version of the line.)

If the ground surface was level and the soil thickness similar, very good records were produced, however hills and valleys presented a problem. Seismic testing is most useful in offshore prospecting as the ocean surface is level, and the ocean bottom is fairly flat or just gently sloped. Explosions and undirected seismic waves have been a problem for sea life, particularly whales who communicate with their own sounds.

Geologists and geophysicists are often at odds with each other. Geophysicists believe their maps are far better than the geologists' maps. Geologists accuse geophysicists of not paying attention to the well data present in the area. It really takes a blend of the two disciplines to form the most comprehensive maps.

Yours truly had just enough geophysics courses and background to be dangerous. Occasionally, in Division Committee meetings, I was critical of a seismic-recommended project which was proposed in my assigned area, however seismic interpretations were viewed as the final word. Seismic lines and maps were considered confidential data and kept in securely locked areas. To look at a line, one had to obtain permission and then sign out the data. Geological data was kept in unlocked file cabinets and often just laid about on our drafting tables.

On one such project, I examined all the geophysical lines and maps, and I believed them to be incorrect following my review. Seismic maps attempted to tie the seismic reflection data to existing wells along the seismic lines. Each well on their map had a small note next to the well location as to the amount of mis-tie of the seismic interpretation at that well spot to the actual well data horizon. I took their map and overlaid it with translucent paper and contoured a map simply consisting of their mis-tie information. One well was right on target, another one was 20' off, another well was 50' off and, the well next to their proposed drilling location was 173' off the target zone. They believed the target anticline had 50' of structural closure. I took my contoured mis-tie map to Division Committee and the deal was instantly killed. Is it any wonder why I was not popular in the Geophysical Department?

However, my great wildcat discovery of South Calumet Field—with over 300 billion cubic feet of gas reserves—was located on my interpretation of two old seismic lines. With only two wells in the township, I interpreted a thick area of probable sand in the center, as the two seismic lines indicated a syncline through the middle of the township, and I could see a thickening of the target section in the syncline. The geophysicists thought I was crazy and it would be a dry hole for sure. They did concede however that my interpretation was correct, and thickening was indicated on the seismic lines. Lucky for me, they were not making the decision to drill the well, and I convinced management to go ahead with it.

Besides looking for structures, geophysics has been and is currently being used to see hydrocarbons in place. Gas and oil reflect seismic waves differently from saltwater reflections. Seismic analysis permits us to see these very minor changes, and gives us another tool for finding oil and gas. The point we should not forget is that the seismic reflections all depend upon the rocks. A slight change in lithology (the general physical characteristics of rocks) may affect the reflections more than the presence of the fluids within the rocks.

This map was made maybe five years after the one on page 34.

It shows the South Calumet Field has grown to 15 sections in area and has merged with the NW Calumet Field in 14N-9W. The productive sands were all in the Atoka-Morrow section, but were a series of at least 10 different sands. In later years, the field was infill drilled and redrilled with approximately 150 wells.

At the time this map was made the SW Altona Field had expanded to eight sections. It also had many more wells drilled, but many of the sands were water bearing.

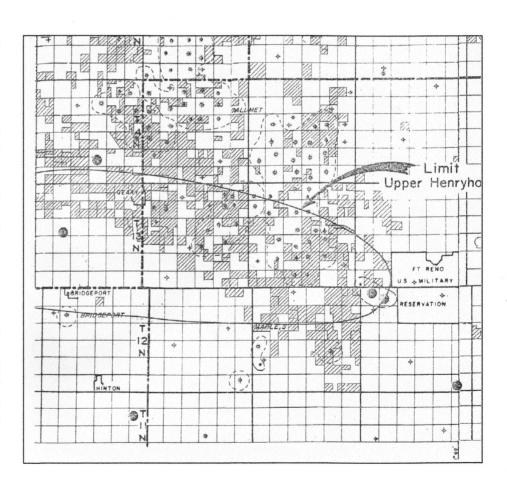

◆ CHAPTER 21 ◆ The South Vici oil & gas prospect

How to spend 25 years of your life chasing a stratigraphic trap — is there really a pot of gold at the end of the rainbow?

Most of my career I spent chasing stratigraphic traps in the Morrow and Springer Formations of Pennsylvanian Age. My major wildcat discovery was the South Calumet Field, which I found for Pan Am in 1966. It was the very first wildcat prospect I had originated and the best wildcat discovery of my career. Over 150 wells were subsequently drilled in the township, resulting in Pan Am finding over 300 billion cubic feet of gas, costing the industry maybe $75 million dollars — and returning three to five times profit on their investment. I never got to participate in any of these wells. (In retrospect, I should have left Pan Am right after the discovery, but it was just too early in my career.)

The Morrow sands stretch across the Anadarko Basin from central Oklahoma into the Texas Panhandle, a span of over 300 miles. The updip northern edge sands pinch out and onlap along an unconformity, and vary drastically from well to offset well, from thin, tight sands to sands 50'+ in thickness. Geologically, they map as a series of beach sands, thin offshore bars, and surge channel sands cutting and reworking the beaches and bars. I put together several deals along the updip edge of the Morrow limit. These were all offset development wells. My overriding royalty income from these wells was inversely proportional to the reserves in the well — the larger my interest, the poorer the well seemed to perform had been my luck.

The best Morrow sands were found as thicker, more easily mapped offshore bar sands — like a modern Galveston Island. The best field in the trend was named Lenora Field. It's a couple miles wide and about eight miles long, with reserves of over 300 BCFG. The industry chased this sand to the southeast and the northwest, but never found another Lenora Field. Of course I had a great prospect on the sand trend, and my ego assured me I also knew I was smarter than the industry!

In early 1978, my friend Ron Eddington (remember, I got his office at Pan Am), came to see me with a lease he had just purchased as he needed a partner to cover the cost. A recent well had been completed in the Red Fork Formation in an adjoining section in Dewey County, Oklahoma. I whipped out one of my maps and said, "I think the Lenora Sands trend through this area, and we could be looking at some really good production. Let's put a larger deal together for a wildcat with offset production from the Red Fork." (It was always easier to sell offset deals over wildcats.)

We had the one 160-acre lease, but needed lots more. Ron said he heard from a promoter named Jack Smith (maybe an alias) that Excalibur Oil was a possible buyer. They were a new outfit, had fancy offices in Tulsa, and they had just been profiled by Hugh Downs on the TV program 20/20 with their principals flying around the country from deal to deal in their company jet. They also owned an Excalibur special edition automobile which they drove to our home office. When our meeting was over, their fancy car wouldn't start. For some reason, we could not jump start it either, so they had to call a tow truck. (I wonder if that should have served as a premonition of what our relationship was to be?) Their management looked at the deal, decided to take it, and we formed a nine-section Area of Interest in which they agreed to purchase all the open acreage. They put the section together to be drilled on 640-acre spacing for the Morrow sands. They force-pooled open acreage at the Corporation Commission, and we were happy the project was moving along so well.

Excalibur hired Robinson Brothers Drilling Co. to drill the well at our preferred location, however we made the mistake of not insisting on doing the wellsite geology during the drilling. In late 1980, the #1-14 Castor well was drilled. However Robinson Brothers were not good operators and their practice was to cut corners at any opportunity (they had a blowout a couple years later and went bankrupt in a well in which 150% of the interests had been sold.)

They were drilling with basically water, very poor mud, with no weight to prevent blowouts. When they hit the Red Fork Sand, it was 50' thick and loaded with oil and high gas pressure. The well tried to blow out. It was the thickest sand in the area. Casing was set in the bottom 10' of sand. The bottom was squeezed three times with lots of cement to hold the pressure, but continued to leak gas through the rest of the downhole drilling. If Ron or I had been sitting the well, we would have made them drill another 50' before setting casing. The well was drilled deeper below the uphole casing (which was smaller than what we had recommended), through two Lenora type sands, each 50'+ in thickness. The top sand appeared water-free, with gas shows, whereas the bottom sand calculated wet.

At this point, Ron and I were ecstatic with joy and began planning our retirements and exotic trips around the globe! We calculated from 6–20 billion cubic feet of gas reserves and over 200,000 barrels of oil and condensate from just the #1-14 Castor well—without looking at field offsets. Immediately we asked Excalibur for our acreage picture. How much had they purchased? They were vague about their holdings, so we sent our landman into the field to see what they had purchased and what was still available. News travels fast in the oil patch, and landmen quickly gobble up open acreage. We found out that Excalibur had not purchased anything outside of the section as per our agreement. Ron and I soon drafted in $875,000 worth of leases in the nine-section area of interest—and tried to get them to pay for it.

It seemed Excalibur had a Canadian partner, Dalco Oil out of Calgary. Their deal with Excalibur was different than our deal (apparently they were cutting out our interest and increasing theirs). We all met with our lawyers, and the months wore on. The interest rate rose to 22% and the bank was getting all our fortunes (which were not yet actually in-hand, mind you).

In early 1981, Robinson Brothers attempted to complete the well. Several million cubic feet of gas was produced from the upper Morrow sand, but channeled water from the lower sand because of a poor cement job. They set a plug in the bottom of the well and moved up to the Red Fork Sand. Cement from the extensive squeeze jobs had plugged the Red Fork sand at the wellbore. Ron and I told them to be gentle and not give the Red Fork Sand extensive acid or water which reacts with clays in the sand—but of course that's precisely what they did. An emulsion block was formed and very little came out of the well. Seven years later when we took over the well, it still produced emulsified oil! A pump was put on the well and reportedly produced 10,000 barrels of oil before it was plugged. (I finally got a check from the bankruptcy court for $1.50 for this fiasco.) The total well costs had gone from $600,000 to over $1,500,000—not counting legal costs. Lease costs in the area based on the well logs rose to over $1000/acre.

We had all these leases and needed new partners quickly.

We sold acreage in two sections to the east to Natomas Oil, who drilled two wells in thin, upper Morrow sands. They never drilled to the lower sands. I took a small interest in both of the wells to help get them drilled. These finally paid out a few years later.

We sold the bulk of our acreage to Clayton Lee who was building deep gas drilling rigs at $6M each. He and a Texas partner of his were building six rigs. Clayton agreed to pick up the interest at the bank and purchase the leases as soon as he sold one of his new rigs. Clayton then had a heart attack and cut back on his business, and later went into bankruptcy—owing us $1,000,000. We took the leases back and looked for new partners. We could have sued Clayton for breach of contract, but the leases would expire before we got through the courts. Also, Clayton was a friend of ours, and who would sue a friend just trying to recover from a heart attack?

We found a reputable company, Edwards & Leach Oil Co., to take our lease position and drill a new well in section 14. The lawyers said we couldn't get title to the section or get Robinson Bros. out of it. My firm, Cochrane Exploration Corp. (CEC), was a partner in the #1-14 Castor well, and their lawyers claimed that CEC was subject to the operating agreement that named them (Robinson Bros.) as the operator. I never signed the operating agreement and therefore was not legally bound by it. I took them to the Corporation Commission and instead named Edwards & Leach (ELOC) as my operator. The Commission upheld my position, and we got a new well drilled. I wanted to drill updip in the northeast quarter section, but ELOC decided they liked the southwest quarter section better. It made a small well from several horizons. We proposed another well.

ELOC got some new partners in 1982 through a Texas promoter, Don Woody. This guy was an international traveler and promoter. He captured a Lebanese Banker from Beirut, who moved his bank to Crete. He built the bank on the border between the Greek side of Crete and the Arab side. You could walk in one country and exit in the next. He had an American wife. The banker's name was Fareed Saab and he was slick too. He got to me owing $300,000 for leases he never paid for. However, the Texas Promoter got Saab for over a million. Where did the money go? Don Woody told us he had bought some leases with the money, and they expired. (Yeah, sure! I mentally planned a trip to Dallas to break his legs, or more — but my better angels prevailed and I decided this feeling of vengeance wasn't worth landing in prison.)

Time goes on, and old deals never go away. Maybe my geologist self just can't believe I was wrong or that the oil isn't there someplace! In 1986, after not thinking about it for a long time, I interested my then-partner, Bob Marple, in trying to rework the #1-14 Castor — that well that had started the whole fiasco. At the time, Bob operated as Sooner Crude and I operated as Cochrane Exploration. We agree to go 50-50, with Sooner Crude designated to operate the well.

We purchased the well from the bankruptcy court for $25,000. We acquired some of the existing partners' interests in the other two wells in the section. We sold some of the interests to other participants who were working interest partners in other deals of ours. We patched a hole in the casing with cement. (It failed again later though—dumping a lot of frack water and emulsified oil back into the hole.) We gave the formation a diesel oil frack to clean out the perforations. We produced some new oil, but not what we'd hoped. We then went down to the Morrow section which only had a simple plug which was easily removed. The gauge on the wellhead showed over 3200 p.s.i. We opened the Morrow and flowed gas water free for 20 days at one to two million cubic feet of gas per day. The well then channeled water in from the lower zone, which we spent more money trying to fix, but finally gave up. We had spent about $100,000 trying to fix the well. After all this…I still believe that it's there, close to that wellbore!

The story still goes on. It's now 1993, and a new player, Mewbourne Oil out of Texas, buys some of the ELOC interests in two sections. Mewbourne files a spacing and pooling action to redrill Section 14 at the Corporation Commission. We hire lawyers to protect our interests. We then sell out part of our interests to Mewbourne. I decided to join in their first well, the #1-14 Cain. (I wonder if Abel was his brother?) The operation makes a nice well in the Upper Morrow sand, and didn't penetrate to the lower sand like I wanted. Their costs, I thought, were high so I went non-consent on their

Me and Bob Marple (r) at the Vici well—the gauge shows over 3000 psi shut-in pressure. (1986)

second well. It was a pretty good well, but took three years for me to back-in.

One of the wells to the east has now been plugged, and also two wells to the west. Three wells are still barely producing, but soon will be plugged. If I was younger, I no doubt would promote another well to look for the Lenora Sand. When you look at the land under your feet, think how much it changes in a mile, a half mile, or even a few feet! You know, it still might be out there someplace!

Want to drill a well…?

♦ CHAPTER 22 ♦   Oil companies

A little bit of history, and changes in the industry

From the 1890s through the 1920s, the oil industry made many significant discoveries in North American which led to the creation of some giant oil companies. By the time Standard Oil was broken up by Teddy Roosevelt under the Sherman Anti-Trust Act of 1890 (in a battle which lasted from 1909 through 1911), Standard Oil controlled over 80% of virtually all segments of the oil industry.

The founder of Pan American Petroleum, where I worked during my short four-year career with "Big Oil, was Edward L. Doheny. In 1892, Doheny discovered oil in Los Angeles which brought him his first fortune. In 1901, Doheny founded the Mexican Petroleum Company. Then, in Vera Cruz, Mexico, he hit it really big. At 598' depth, he hit a gusher, which soon pumped 260,000 bbls. of oil per day. The well produced 57 million bbls. of oil during its productive lifetime.

In 1916, Doheny formed Pan American Petroleum and Transport Company. By 1921, Pan American was the largest oil company in the United States. The companies spawned by the 1911 breakup of Standard Oil were also growing. These successor companies were:

Standard Oil of New Jersey (now EXXON) with 50% of the assets

Standard Oil of New York (Mobil) with 9% of the assets

Standard Oil of California (Chevron)

Standard Oil of Ohio (Sohio)

Standard Oil of Indiana (Stanolind, later Pan Am, later AMOCO)

Continental Oil (CONOCO)

Atlantic Oil (Sun, later ARCO)

In 1922, the U.S. Secretary of the Interior, Albert B. Fall, leased the Elk Hills, California oil field to Pan American Petroleum and Transport Company. He also leased the Teapot Dome field in Wyoming to Sinclair oil Company. Edward L. Doheny and Harry F. Sinclair had conspired with

Albert Fall to get the leases from the government by skirting competitive bidding. Doheny had loaned Fall $100,000 in cash and Sinclair had paid him around $400,000. This became known as the Teapot Dome Scandal. Doheny actually confessed to a U.S. Senate Committee in 1924 that he had loaned Fall $100,000 expressly for this purpose.

Doheny got off without jail time, but Fall and Sinclair did six months each. Leases were returned to the government and big losses were incurred by Sinclair Oil and Pan American Petroleum. In 1925, Standard of Indiana bought part of Pan American's holdings. In 1954, Pan American merged with Stanolind Oil.

In 1998, the renamed company, AMOCO merged with British Petroleum and is now known as BP.

When I joined Pan American Petroleum in 1964, it was widely considered a friendly and terrific place to work by those in the industry — certainly in comparison to some of the other companies I later became familiar with. In fact, Pan Am trained many people in the gas/petroleum exploration industry. Like me, they lasted three or four years and then moved on to other — usually smaller — companies. Pan Am worked well with ex-employees, many of whom generated deals adjacent to their acreage holdings.

By the 1960s, the industry had morphed from where it was before the Roosevelt-led breakup. Companies were less secretive and collaborated together on many plays. In the old days, oil companies had "oil scouts" who prowled around the oil patch to find out who was buying leases and where. Drilling wells were carefully researched, the depth they drilled determined by counting the pipe pulls of the drill bit, drill stem tests were carefully monitored, etc. Unauthorized persons were not allowed on the rig floor or allowed access to the workers on the rigs.

But scouting became friendlier over time, and daily or weekly meetings occurred in town or at local cafes where some information was thus exchanged. Companies built well files of scouting information. States required permits to drill and kept statistics on well tests and completion horizons and initial potentials (IPs). By my time in the business, all this type of data was fed to an industry-formed business — Petroleum Information. Completion cards (4" x 6") were published daily and sold to the industry. In 1972, I purchased an old set of Scout Tickets from the widow of a geologist who had recently died. I paid $2500 for the set which went back to old handwritten tickets in the early part of the century. I kept these tickets up to date until I retired in the 1980s and computers had taken over the keeping of records by then. I took to the dump (recycled) over 300,000 cards I had acquired through the years, having spent over $50,000 to accumulate them. (That was a sad day!)

Big oil companies operated differently from each other and often didn't like one another. The structure of Pan Am was oriented toward exploration—some companies instead spent most of their money on development—offsetting known production. Companies like Shell, Pan Am, Mobil, Texaco, Chevron, and others bought large numbers of leases in scantily explored basins for real exploratory wildcats. Maybe they purchased 100,000 acres of leases in a regional geologic trend. A standard lease gave the landowner $1/8^{th}$ to $3/16^{th}$ of the hoped-for future income from the well to be drilled—the company took the rest. In "dead" areas, leases were cheap—$5 to $10 per acre. Terms were often for 10 years—sometimes fewer. Once a wildcat had struck oil, the terms and cost rose drastically. At $10 per acre, a whole section only costs $6400. A wildcat discovery could then easily push the price to a $1000/acre. Not knowing where the field might develop, the companies wanted to cover larger areas, so they bought 160 acres in a section and left the rest. Other companies often did the same thing, so a proposed well often had other industry partners. Pan Am's philosophy was that to drill a wildcat, the company had to own over 50% of a nine-section area (for a 640-acre spaced gas well—oil wells are spaced on smaller land units depending upon the depth of the target horizon.)

If the wildcats were not economically successful or potentially had large reserves, the company(s) would farmout their acreage to other companies. Farmouts also made the company money without the large expense of drilling the wells. Here's a typical term for a farmout: the company provided the lease to drill on, the landowner still had their 1/8 royalty, and the company added another 1/8 overriding interest (ORRI) which gave the company income from producing wells. This reduced the net revenue in the production to 75% for the company taking the farmout and covered 100% of the costs for the well. We poor independent consulting geologists who might've brokered the deal to find investors were lucky to put another 1% ORRI on the deal. Sometimes we took a back-in after payout, which meant the company taking the deal got their investment returned 100% and then we poor geologists came back-in for a percentage of the production. The best deal I ever got was a 25% back-in, whereas usually it was 3 or 4 percent. By the time we backed into the well, the flush production was gone and the well was only producing a few barrels of oil per day. Once I backed into a well 10 years after it had been drilled.

Big oil companies did not give ORRIs to their employees. Our salaries were supposed to be the highest in the industry. Small companies gave ORRIs to their employees as an additional incentive to instead work for them.

◆ CHAPTER 23 ◆ Other oil companies

Do creative people need isolation or competitive interaction to perform well?

The following discussion of other oil companies are a few observations gathered through the years, mostly after I had left Pan American. The interactions between employees of major companies is always quite professional during business hours, but can be much looser at social events. Whereas consultants, I think, have more fun dealing with major oil companies—at least I did. There's always the interplay of employees and management in major oil companies. One must appear to be professional and spout the company line as needed. Consultants need something from the major oil company—a piece of data, an electric log, a farmout of a lease, or the company to join in the drilling of a particular well. The companies need consultants and small oil companies to take farmouts and prove up their acreage holdings. Being a fun-loving "good ole' boy" breaks the ice. What is the most recent joke that you heard? The important thing however is to have a track record of getting things done..

Pan American Petroleum was a friendly place to work, with a good exchange of ideas among its employees. We socialized with each other—geologists, landmen, and engineers. There was some rivalry between geologists, geophysicists, and engineers, but it was mostly low-key. Ex-employees remained old friends and tended to stay in touch, at least to some degree.

Philips Oil Company was one of the different ones though. While working on one farmout deal, I spent the better part of a week in their office reviewing seismic records. The office had a large and centrally located open file room surrounded by very small offices. Two secretaries had a desk on each end of the file room. The geologists and geophysicists came to work, entered their offices, and closed their doors. All phone calls came in via the secretaries and were forwarded to the appropriate office. Apparently, they could make outgoing calls without going through a secretary. If a geologist or geophysicist wanted a file or some electric logs or maps, he phoned the secretary and she then searched the files and took the materials into their office for them. I asked about the setup as it appeared very sterile and unfriendly. Their philosophy was that these people were creative and needed the quiet of their closed office in order to produce.

That certainly was different from my philosophy. Mine was read all the files, talk with the authors, and pull out any ideas that he (they) might have had which were not included in the official report. I also talked with the operations geologists who actually drilled the wells recommended in the reports as I was keen to find out what they had learned. Since it was all new to me, I thought it simply made sense to start where they left off. But overall, to me, the key was interaction with my colleagues. After all, why reinvent the wheel?

Texaco had a reputation as a more secretive company and many people did not like them. One geologist I knew when asked who he worked for often replied, "I work for a major oil company." He rarely acknowledged he worked for Texaco. I never went to their offices.

Texas Oil & Gas was a different outfit, very fun-loving. They bought acreage next to other companies' wildcats and good development wells. They did not drill wildcat wells. If it was a 640-acre unit, they drilled as close as they could legally get to the producing offset well. Sometimes they got an exception to the normal spacing and drilled closer to the producing well. Their excuse to the Corporation Commission for the exception was there was a stream there, or a farmhouse there, or a road, or whatever. Sometimes it worked and they secured a permit to drill off pattern. If the offset well was owned by a major oil company or someone who didn't like them, it was protested. Occasionally the Corporation Commission allowed the offset well to be drilled along with a penalty being assessed—so they could only produce the well at a lower rate than the basic allowable allocation.

Texas Oil & Gas hired an unusually large number of secretaries—more than they needed for a normal operation. The girls were always pretty, lighthearted, and regularly enjoyed drinking and partying. TOG geologists and landmen always took two or three of the girls out to lunch each day. They would meet up with other oilmen and a party soon commenced. The girls ultimately gathered information for the company, so the company profited from all this social interaction.

During some of the boom days, when everybody had a deal or was looking for a deal, clients would come to town, stop in at the Davis, Northcutt and Cochrane office and want to look at deals. It seemed like this was always near lunchtime so the subtext indicated the client simply wanted to party. I would call Texas Oil & Gas, tell them Earl was in town and we wanted to show him deals in Blaine & Canadian Counties. I'd ask, "Have you got three or four girls who can meet us for lunch?" They always provided, especially if they owned leases near our prospects. Sometimes the girls got back to their offices by quitting time. Sometimes we were still partying at midnight! For fear of alcoholism, I gave up any drinking on weekends and non-client days!

Skelly Oil was a staid old gas company. I often met with their manager, who truly was an alcoholic. Over the years, I sold them a couple of small deals. My main contact there was with a company manager named Jay—I was in my 30s while he was near retirement age. Our "meetings" always occurred at the bar where we would sit and drink scotch. Jay was a short, pudgy guy with an oddly pasty complexion who never appeared drunk. His speech was normally a slow drawl, so there was no apparent change from the alcohol. Smack in the middle of a sentence, he would suddenly fall asleep while leaning on the bar. But somehow he never fell off his barstool—his usual pattern was to sleep about 15 minutes and then wake up and order another drink. I learned to stop imbibing while he was asleep and instead would get myself a cup of coffee, hit the restroom, or eat a snack. Once he'd revived, our discussion and drinking would then continue until quitting time when he'd head for home.

Kerr-McGee Oil Company's home office was in Oklahoma City. Kerr was Senator Kerr, a longtime presence in Washington, D.C., McGee was a paranoid geologist. Kerr-McGee came into existence during WW II. McGee had drilled productive wells in Louisiana right up to the water's edge. Obviously, there was oil under the bay extending the field. Senator Kerr told Congress we needed the oil for national defense purposes and that barges could be constructed to drill offshore. However, this would be expensive, so he also said the government should help with the cost to develop offshore rigs. The government duly paid Brown and Root to build the first offshore platform to drill the first oil field developed "out of sight of land." The first well in the field was drilled in October 1947. Kerr-McGee became rich and a new phase of offshore drilling developed on the heels of this initial project. Brown and Root had been around for many years and developed many construction projects for the government and the military. Brown & Root later became Haliburton, and more recently split off their construction arm, now called KBR, an entity actively working in the Middle East for the US military.

I made friends with Rich McKinney, the exploration manager for Kerr-McGee. I showed him several deals through the years, but Dean McGee never liked any of them. Security was the name of the game in their office. There was one locked door on the street into the office, with two armed guards at the desk. You rang, one of them let you in. They gave you a once-over and looked into your briefcase. They then called McKinney, and he came down in the elevator and took me back to his office or to a conference room. He often had another geologist, a landman, or geophysicist attend these meetings. After going through this process several times, I called McKinney with another deal. He said, "Bring it over and we'll look at it." I then replied, "No, come to my office. We won't search you and will even offer you a drink. What do you like?" He came to my office. They still never took a deal but we had more fun without me being subjected to their off-putting security measures.

My neighbor Jeff was a Kerr-McGee geochemist who worked in their research lab outside of town. Jeff knew nothing of the oil industry or anything about Kerr-McGee's background. He worked in geochemical research and could not talk about anything the company was doing.

Several years later, Bob Marple and I put a deal together, directly across the river from the Kerr-McGee nuclear facility. They had drilled a 2000' deep nuclear disposal well across from our prospect. I was worried we might drill into a crack and suddenly have nuclear waste coming out of our well. So I bought a Geiger counter and we monitored the mud from the surface to total depth. Thankfully we encountered no radioactive materials. Kerr-McGee released no data on what they had put in the well. I asked my neighbor and Rich McKinney to research their records, but they were nonexistent—or classified. I asked the Oklahoma Corporation Commission to get the information from Kerr-McGee, but they never received anything. (Remember the infamous Karen Silkwood incident?)

Selling prospects to these companies

One criteria I had when I sold a deal to a company was that I would get the well-site geology work and supervise the running of the electric logs so as to give my recommendations for possible testing. Since I received an ORRI (Over Riding Royalty Interest) on the project, my input was valuable for my future income.

One time I was on one of Anadarko's wells, it was Saturday with the prospective oil horizon pay zone being cut and looking even better than when I had mapped it. I called the exploration manager in charge of the office and told him the news. It looked like we might log the well that night. What did Anadarko want to do? He said, "Oh, bring the logs into the office Monday morning by 10:00 a.m. and we'll look at them. Just have the rig circulate the mud until we make a decision." He further said, based on my enthusiasm, "Make sure we run a dipmeter so we can possibly figure out where the sand trend goes. I'll call a geophysical company to run out from town and shoot a seismic line through the well and along your trend."

I couldn't believe he could work so fast and he really was enthusiastic. The seismic crew actually showed up in the early afternoon. Ah, there was a glitch. The farmer said, "You can run the seismic survey across my land for $2,500, but not until you pay me." I called the exploration manager to bring someone out with the money. He said "No one in the company works on weekends and I'm not going to run out there. Can you pay them? We can reimburse you on Monday morning." Well, I didn't have $2500 in my checking account, but luckily I had a check in my wallet—so I paid the farmer and the seismic crew did their work. They were parked there on standby time of $600 per hour, so I just saved the company more money.

I think you can get a sense of the pressure we were always under with so many different projects going on—not to mention the pressure of relationships and raising families (I think my kids basically raised themselves). Since the average deal took a year to put together, and then maybe another year to actually get it drilled, many deals were at some stage of development at the same time. I was constantly looking for new projects while concurrently starting or continuing ongoing studies. There was also always a client who needed to talk about on offset well, or a different zone in a wellbore to rework.

There were draftsmen to supervise, landmen to hire, drilling rigs to be lined up for a new project, new investors to talk to, title lawyers to confer with, lease titles to get cured, companies to contact for new farmouts…and on and on. There were professional meetings to attend, social functions, luncheons, and drinks after work. Work could actually go on 24 hours per day: a well was going to drill the pay zone at midnight, so a quick trip to a drilling well was needed—but maybe I didn't get home till two or three days later.

We lived on coffee, booze, and cigarettes. A coffee pot was going continually and replenished every hour. It was always strong, black, and often tasted burnt. I have this habit of counting things. I count cars, people, birds—lots of things. I averaged 14 cups of coffee (or more) per day plus bottles of antacids (those I didn't keep track of).

Then we all had lots of drinks with clients—everyone. I learned to pace my drinking to one per hour, which is about the rate your body can burn the alcohol. Somehow, I did not become an alcoholic! I virtually never drank at home or on weekends, unless we were having a party or attending one.

Cigarette smokers were virtually everyone and smoking everywhere. I was one of the rare people who did not smoke. However, I was exposed to second-hand smoke everywhere I went—offices, clubs, bars/restaurants, at home, in cars. There were cigarette butts everywhere—too many to count! I did bitch at the number of ashtrays around the house—many with smoldering cigarette butts.

Ann began smoking at age 16 and rapidly developed a three or more packs per day habit. She died in her mid-60s, no doubt asking for another cigarette. My partner, Bob Northcutt was also a chain smoker. Bob had at least a dozen scars on his chest from falling asleep with a cigarette in his mouth. He never burned down a house or a motel, so I guess he was lucky.

When an addicted smoker is out of cigarettes, it typically becomes a real panic for them. There was an all-night Quik Shop about a mile from our house. Ann would shake me awake when she was out of cigarettes and insist I fetch some for her immediately. I would have to get dressed and drive out to do her bidding and she would whine, "You wouldn't want to send a poor girl out in the middle of the night. It's cold, and might be dangerous!" I would return with a carton of cigarettes but also soon learned to hide a couple of packs away someplace to save me a middle-of-the-night run for her. Sometimes Ann would find them and again I'd still have to go out for more.

While I fell asleep early from our fast pace of living, she would read and smoke in bed until late into the night.

Another passion in Oklahoma, Texas (and elsewhere) was and still is football. This often begins in Little League and parents put their kids on teams and attend every game. Somehow, my boys weren't into football, but did play for a time. Passions are high between the teams and parents put undue pressures on their kids. Name calling and even fights among parents happened too often. More than once, I chewed out a parent for their actions. "Lighten up, fellow or ma'am, it's just a game. These are kids and they make mistakes. Get off their butts."

High school football was worse.

Everyone had their favorite college team. Oklahoma has two great college football schools, Oklahoma University and Oklahoma State University. Their rivalry is intense. Then there are the rival out of state teams—notably from Texas and Nebraska. One time my sister and brother-in-law visited from Nebraska and we went to the O.U. Nebraska game. Larry came wearing a Nebraska sweatshirt. Nebraska was winning, and Larry was yelling at every score. There was dead silence around us and I feared we would soon be involved in a brawl. Thankfully O.U. scored and finally managed to just beat Nebraska. I gave up going to games after that.

Then of course, there's professional football, and everyone has their favorite team. Bets were high at all the bars and offices. The problem is that Oklahoma doesn't have a professional football team. Oklahomans don't like Texas, so it's difficult to yell for the Cowboys. But then Barry Switzer, the O.U. coach went to coach the Cowboys and people started to root for the Cowboys as Barry was well liked in Oklahoma. His salary was capped at O.U. to some kind of "reasonable" limit. The Alumni took care of him. Barry was hired for all kinds of TV ads. His face popped up continually everywhere, on billboards, and virtually on every other TV commercial.

Tom Landry, the previous Cowboy coach in Texas showed up on TV selling bibles and presenting a very pious persona. On TV during the Cowboy games, one could read his lips as he cussed out the refs or the other team. Pious indeed!

Chasing tornados

We are back in western Oklahoma during the summer tornado season. I had been sent out to a drilling well to look at a potential pay zone. The well is drilling slowly and there are several hours before the zone will be cut. Ah, what to do? I left the book I was reading back in the city. Damn! The nearest town with a good café is 25 miles away. The radio says a tornado has touched down on the ground just 15 miles north of the rig. Egad! I think I'll go chase it for a closer look.

The area contains only dirt roads laid out on a half-mile grid pattern. The farms and houses are far between. Often, we (I) tear down these straight roads at 60–70 mph. There are no stop signs — or they're ignored. You can see the dust plume from a car or truck on an adjacent road more than a mile away. The goal in these parts was to adjust your speed and reach an intersection well ahead of or behind other cars — of which there are few hereabout.

The radio now reports the tornado has touched down about 10 miles east of its last-known location. I quickly turn and shoot off to the east, pedal to the metal. Ten minutes later, I approach a black cloud with a protrusion beneath it, which is the tornado itself not quite touching the ground. It doesn't look like it's descending toward the ground, so I speed on toward it. By the time I get there, the tornado has withdrawn up into the clouds.

Well, I decide maybe I'll now head to town for coffee or a beer and listen for the next weather update. Maybe next time I'll get a better look!

Tornado-watching during tornado season is a must and can also save your life. People keep their radios and TVs on and weather reports are given throughout the day. Sirens go off in the cities and towns when tornados are sighted. Many people in this region have tornado shelters, and we previously had one too at our home on Thompson Avenue in Oklahoma City which was built of concrete. In 15 years we only went into it twice though—the spiders got more use out of it than we did!

The southwest part of Oklahoma City (Moore, Oklahoma) has been hit several times with horrendous destruction by violent twisters. We've lived in the northeast and northwest parts of the city, which have missed most of the tornado touchdowns. My theory is when they cross Lake Hefner, they withdraw into the clouds before touching down further to the northeast.

In 1969, when my son, Brian, was in preschool at a Jewish Temple (albeit an odd spot for a good Catholic kid), a tornado touched down during the weekend on Pennsylvania Avenue when the children weren't there. It took off the top of the two-story brick building, everything was sucked out of the second-floor classroom—furniture and all. We went by to see the damage and I looked down into the bushes near the street. I found a kid's drawing in perfect condition. It was signed in the lower corner in crayon, "Brian!" I wonder if his mother still has the drawing? I must ask her.

If you tune into the Weather Channel, especially during tornado season, you will meet the current bunch of crazy tornado chasers. Besides obtaining astounding pictures of these dramatic events, they have also measured and mapped wind speeds within the tornados themselves—however their daring escapades have cost some their lives.

Rattlesnakes

Watch out for the rattlesnakes, Oklahoma has lots of them! They semi-hibernate in the wintertime, and always come out in springtime. There's a line of gypsum hills which run north-south in western Oklahoma. Two or three towns are located adjacent to these hills. The snakes love these gypsum hills as there are lots of friendly crevices in the brittle rocks located there.

Each spring, the local towns have "Rattlesnake Days" and people flock out from the surrounding cities to see the fun, and we always went and took our children along with us. The snake hunters are all there in full force. The snakes themselves, the stars of the show, are kind of sluggish when they first awaken, so they're easy to catch. (I never thought I wanted to participate in the catching myself though!)

The snake hunters bring all the rattlesnakes back to town and dump them in pits setup on the street, and there are prizes for the largest snakes caught. Soon, there are hundreds of snakes squirming and writhing in these pits, and one of the brave snake hunters climbs into the enclosure and picks up snakes to show to the public — particularly to the young children, who squeal and yell in a mixture of terror and delight.

Snakes are milked for their venom. They are killed, cut up, and barbequed in Oklahoma. Of course, we have to eat them, plus also offer bites of snake to the children. The braver ones have a taste. Once you put barbeque sauce on them, who knows what they taste like? So, after this great treat, we then have a picnic with real food which we brought with us — drink a couple of beers, and vow to come back again next year.

I sat on many wells in rattlesnake country. Drilling rigs need space for equipment and parking. Usually about an acre of land is scraped clean for the rig and equipment. One night, I had to go check a rig for a potential pay zone. I arrived at the rig and saw that the cleared area is two or three times as large as most rigs require. Maybe 20 dead rattlesnakes are hung on the fence along the road into the well site. I drive in a circle around the rig. Rattlesnakes are coiled up everywhere, especially close to the drilling rig. The crew thought the snakes liked the vibrations from the drilling engines. No one used the port-a-potty unless they checked around everywhere for snakes first. Two workers would go together out to the mud pit to mix more mud chemicals. The samples I came to look at were supposed to be caught every 10′ of new drilling depth. I think the bags lied about where they were collected. They probably filled all the bags in the morning at one particular drilling depth given their squirmy working conditions there.

What a wonderful place to dwell — tornados, rattlesnakes, dry dust and heat, no trees, a little brush, and wind — year 'round always lots of wind! What is that line in the musical 'Oklahoma'? "... where the wind comes sweeping down the plain."

Cowboy bars

Alcohol-laced good times and dancing too on Saturday nights.

Outside of Oklahoma City, Tulsa, and a few other large towns and cities, Oklahoma is still dry, although it has loosened up in recent years as the suburbs creep further and further from the main cities. However, liquor, wine, and beer can still only be purchased in state licensed liquor stores.

We moved to Oklahoma in 1964, and the state had just gone wet a handful of years before, in 1959. My neighbors wished it was still dry as their bootleggers got them booze anytime, day or night—and with more selection and cheaper prices than what was offered in the state-sanctioned liquor stores.

Bars opened in Oklahoma City soon after the state went wet. They were called 'clubs' and you had to buy a club card ranging from $10 to $100, depending on their exclusivity. I never bought a club card, but always had access to any club I wanted to drink in. When I showed up without a club card, they would invite me in as a guest, but never insisted I purchase a card. The next rule emplaced was that you had to have your own bottle at the club—with your name on it. We brought bottles for a short while and then the "clubs" provided "your bottle" for you with your name on it, and charged you on a per-drink basis. I guess if you didn't come in for a while, your bottle got another name placed on it. After some time, the labels literally just fell off the bottles and that rule was no longer enforced.

My children, when they reached 18 (although some earlier, having lied about their age), worked as waiters, bartenders, or in the kitchen of many of these local clubs. Periodically, the ABC Board or the police raided them to see if laws were being broken. As a result, my children have been arrested, handcuffed, and taken off to jail. The club owner was called each time and he always came down and bailed out his employees and paid a fine. After a while, a phone call warned the employees that a raid was coming at 9:00 p.m., and everything was hurriedly put in order. I wonder what the payoff was?

However, the small towns in western Oklahoma were different. There were no clubs, but beer bars with only 3.2% alcohol level beer in many towns (average beer alcohol is 4%–6%). Watonga was one of the spots closest to the wells I was working on, so off to its bars we went. The My-Oh-My Bar was a favorite and was located in an old hotel. When you sat at the bar next to a cast iron pipe which came down from upstairs, you'd hear swoosh—and you could tell someone had just flushed the toilet. The jukebox was loud and blaring a steady stream of cowboy and western songs. On weekends, there was a small band and lots of fun. People would come dancing off the floor and then quickly return, singing the next song along with the band. We laughed a lot!

There were always fights in these bars. Someone was messing with someone else's girlfriend or wife—you wanted to keep your wits about you and not get smacked with a pool cue! The local sheriffs rarely came into the bars and rarely gave out any tickets for speeding. Occasionally they had to arrest somebody—who was invariably one of their neighbors. I don't remember a lot of auto accidents, but many of us got behind the wheel after a night of drinking who, if tested, were likely to have been legally drunk.

One of the pushers (drill rig superintendent) on a rig we used often was an Indian known as Chief Haigler. He could drink more beer and straight whiskey without getting noticeably drunk than anyone I ever imbibed with. The oil service companies always had a case of hard liquor in their trunk and readily gave these to their clients. Crown Royal and Johnny Walker Red were their favorites though—not mine. Everyone in the oil patch seemed to drive a dented pickup truck.

No one drank on the rig floors though as they're dangerous places. I'm sure Chief Haigler and I were inebriated many times when we returned to the rig. We always quit drinking for at least an hour and had breakfast with lots of coffee before returning to the rig.

(My son, Chris, now owns and operates three sports bars in the area: in Oklahoma City, Edmond, and Moore. I convinced him to have breathalyzers in his bars so people who are over the limit don't drive away. A growing awareness of the resulting dangers has led to groups of partiers now often having designated drivers who don't drink and can ensure everyone gets home safely.)

This vantage point is adjacent to California's Scenic Highway 1 near The Sea Ranch.

◆ CHAPTER 25 ◆ R n' R—and finding paradise at land's end

Turns out helping a friend can change your whole life!

The frenetic pace of the oil business spurred the need to find time to recharge our batteries—which meant enjoying some rest and relaxation, seeing some new sights. A couple of years, we headed to Europe for three weeks or so. When we vacationed nearer home, we took the children on canoeing trips in eastern Oklahoma, and also went camping and hiking in Colorado, Arizona, and New Mexico. We also took some wintertime ski trips in Colorado. In order to let my gray cells cool off, I found that, at times, I just needed to be alone and sit atop a mountain by myself. Colorado and New Mexico were close enough to get away for three or four days.

Circa 1976, in the office next to ours, a retired Mobil geologist, Suzanne Takken set up consulting. I had known her for many years while she worked at Mobil. Now she was retired and wanted to do some consulting and try to develop a favorite project. While settling in, she decided she wanted her new office decorated with some wallpaper, however the building's management wanted an exorbitant amount to do the remodel and was not interested in providing wallpaper. I volunteered to do the job as I had learned how to hang wallpaper from my grandmother. I told her, "Suzanne, you mix the drinks, and I can do it in a couple of hours on Saturday morning." She took me up on my offer and we had fun handling the task as she pitched in, and she loved the results.

Neither Ann nor I had been to California. My dad had spent most of his life there, but I had never visited him. I wanted to see the golden state and especially the redwoods. At this point, I had already been in all the states except for Washington, Oregon, California, Utah, Idaho, Hawaii, and Nevada. My father had been born in Washington and spent most of his life in California. So our plan was to take a long car trip, maybe for three or four weeks, to the Tetons and Yellowstone, on into Idaho, catch a little of southern Washington, then head south through Oregon and on into California through the Avenue of the Giants, and then further south in California, before heading back to Oklahoma on I-40.

I told Suzanne of our forthcoming trip to the west coast and how I was arriving in California from the north and driving down through the 'Avenue of the Giants.' She handed us a key and said,

"Why don't you get some real R & R and spend a week at my cabin at The Sea Ranch in Northern California?"

So, three weeks later, after falling in love with the redwoods while touring Avenue of the Giants, we arrived at The Sea Ranch, situated at the northernmost tip of Sonoma County and featuring a spectacular 10-mile stretch of the Northern California coast. The redwood forest stretched inland for several miles and came down out of the coastal hills, meeting scenic Highway One which bisects The Sea Ranch one-half mile east of the ocean.

Drawn as we were to the inviting forests, we immediately fell in love with this secluded coastal area. It was such a contrast from the flat dryness of Oklahoma and Kansas. Instead it was similar to the topography of upstate New York, but with a gorgeous coastline and soaring redwoods—so this landscape rich in natural beauty immediately made me feel "at home."

We stopped at the Lodge for a drink and snagged a salesman to show us the Ranch. That evening, Suzanne called, "Have you bought a lot yet?" I reply, "Well, I did put an offer on a forest lot with real redwood trees. I didn't know people could actually own them. I thought they were all in parks or owned by timber companies!"

I guess she knew I would be hooked. She and I subsequently spent many hours back in Oklahoma City discussing the geology of the west coast, and so my education of this stretch of coastline began.

We spent the week exploring the area. Since 1976, I have never spent less than a month a year at The Sea Ranch, and I moved here full-time in 1988.

Like others who've made this special place their home, I too have come to believe we don't choose The Sea Ranch, The Sea Ranch chooses us—it's truly a marvelous place!

Mine and Ann's first Sea Ranch house (right, front). There are 18 clustered houses only 10' apart in this group which sit at an elevation of 295'.

All have views and ours was in the front row overlooking the ocean.

Our second Sea Ranch place, (left), the beach house, was built by our family in 1986. We took more than six months off work and spent 12 hours per day on its construction, while our two hired hands only worked 8 hours per day.

It's situated on a point at an elevation of just 45'.

Houses on the bluff, The Sea Ranch.

A good place to stroll: Walk On Beach, The Sea Ranch.

◆ CHAPTER 26 ◆   The end of my oil days

How does one get wildcatting out of their blood?

The answer for yours truly is…maybe never! I've invested in several oil deals through the years and some since I retired at The Sea Ranch. Overall, I've had better luck with my own deals. Maybe I had the misfortune to discover the "really big one" with my first wildcat for Pan Am, the largest gas field I would find in my career on my very first try. I wonder… was that good luck or bad luck?

We continued to put oil deals together through the mid-1980s. The deals grew bigger and more expensive over time—and, as a result, more difficult to sell. We got stuck with the $330,000 interest in the Laverty Field deal. We had to mortgage all our properties to pay that one off, but three years later, we recovered the monies in a redrill in the field. The final big one was the Vici deal in which Ron and I drafted in the aforementioned $875,000 worth of leases and, with interest (22% during the Carter years), rose to $1,000,000 of liability.

I drilled several small wildcat discoveries through the years, but they were never of the same caliber as that first one. It would have been nice to have struck it big in the Vici deal, a result which would have ended my career with a feather in my cap. I can still hear the Leprechauns laughing, "Haha! We hid it really good and we're keeping all the gooey black gold for ourselves!" When we finally sold it for much less than break-even monies, we decided we needed to spend much more time at The Sea Ranch.

Ultimately, I'm glad to have gotten out of the oil patch, especially as I felt I never really fit in. I didn't learn to speak the drawl, y'hear! I never bought a cowboy hat or a pair of cowboy boots (or spurs for that matter!)—how do they walk in those things? I never learned to gamble—like they do. Many of them are addicted to gambling and bet on everything, certainly every game played by an Oklahoma or Texas team—in any sport. As a result, Las Vegas is their favorite vacation spot! I never bet with them though, and look at gambling as a statistical game where you cannot win.

I did have one main gamble during my oil patch days though—I gambled on myself. I gambled I could draft hundreds of thousands of dollars of leases into the bank on a 30-day draft and sell them before I had to pay for them. I never turned a lease back for non-payment. I took a working interest in most of the well deals I sold so as to materially participate in the risk, and put most or all of my front-end profit into the drilling of the first well. This made the deals easier to sell and convinced the client I was willing to take the risk alongside them in hopes the proposed well would be successful.

I can't seem to get wildcatting out of my blood though. I purchased the first house on The Sea Ranch as a place to relax, but then we began to think of living there full-time and retirement. So this led to the construction of "the beach house" and spending more time on the coast.

Next I got a real estate license, basically so I could justify sitting around the Lodge chatting with my real estate friends. At this point, the newly-formed California Coastal Commission had reduced the planned size of The Sea Ranch from 5000 lots to 2200 lots, and the sale of new lots ceased, and the developer was divesting their remaining assets. So I put a group of oil investors, partners, and my Sea Ranch real estate pals all together to purchase The Sea Ranch Lodge and the 9-hole golf course from the departing developer, Castle & Cooke. The 5000 acre sheep ranch was reduced to 3300 acres by the sale of three patches of forest on the east side of Highway One.

At that point, the Lodge had only 20 lodging units, a store, post office, and restaurant sitting on 40 acres and would be allowed to expand to 100 lodging units. The golf course has sufficient land to expand to 18 holes. It was a different sort of "wildcat deal" in the making, but regulations and politics control everything and well-intentioned progress soon ground to a snail's pace. The developer made us a "good deal" for a reduced price—but with one little addition. We had to finish building out 45 units of affordable employee housing. There were 15 already constructed, so we had a model for the unit already designed and the acreage set aside for it.

We were able to get the 31 units constructed and sold-out the units for low-income housing and tax credits—which nearly paid for the construction itself. The Sea Ranch Association pulled us into a lawsuit with the developer and our Lodge expansion plans then morphed into a political battle. It took us a couple of years to form an agreement with the The Sea Ranch Association and, by then, the economic climate had changed. We ultimately sold off the Lodge to another entity.

The expansion of the golf course was a completely different situation. The state mandated we could not use potable water to irrigate the golf course. Part of The Sea Ranch is on a sewer system and treats the wastewater to a secondary level. We could increase the treatment level and provide water for the course, but there would not be sufficient water for many years to meet our needs. However luck was with us as just adjacent to The Sea Ranch and the golf course is the small town of Gualala. The town had been shut down for five years with a moratorium on new construction as their old septic leach fields were leaking into the Gualala River. They had a new plan for a treatment plant with the disposal of the secondary treated water into the adjacent redwood forest. But yours truly was the only one authoring an opposing letter to the EPA who needed to approve the project, and I told them "We need this water for our expanded golf course." They concurred and told Gualala to redesign their treatment plan and sell the golf course the water. All well and good for us and we were happy. One major snag: Gualala is in Mendocino County and The Sea Ranch is in Sonoma County. Of course, each of these counties have different boards overseeing sewage and water treatment, not to mention separate building and planning departments. I got to know the Supervisors in both counties. It took a year to redesign the project and get through the political minefield among many sometimes-competing players. But we were ultimately successful in our goal of expanding the golf course.

When we first purchased the Lodge in 1988, Bob and I met a writer working on a historical book which also highlighted local businesses along the coast, Susan Clark. On my part, it was love at first sight. I was immediately struck by her beauty—she's of medium height, has reddish-brown hair with streaks of blonde, freckles, and a very happy, expressive face. And I had also heard good things about her from her best friend, Janann Strand, for maybe the past eight years. In just the preceding three years, Susan had published several short historical articles in The Sea Ranch Association's quarterly publication. I had known Janan for more than a decade as she had a walk-in cabin near our Cluster house. Those cabins were very small, 20′ x 20′ squares with a bedroom loft. Janan had told me about Susan numerous times over the years and kept trying to introduce us as she wanted me to see her maps. But Susan was only at The Sea Ranch part-time and we kept missing the opportunity to meet up.

Our new group of Sea Ranch Lodge owners hired Susan to write a short article on the Lodge for a book she was working on at the time. I attended that meeting and listened to her presentation, possibly asking one or two questions. She has always claimed I intimidated her at that meeting, but our recording of the meeting indicates differently. I think I just intensely stared at her and possibly she was reading my mind. I was instantly attracted to her, initially for her historical knowledge and collection of maps, but surely for her looks and personality too.

She does indeed have a marvelous assortment of maps. Like the fabled Scheherazade, even today she occasionally brings out a new specimen from her collection saying, "Have I not shown you this one?" After that first meeting I pursued her via every possibly excuse...and soon we were in love.

Susan is an architectural historian and was just starting her own business when we finally met and soon I began accompanying her to visit projects she was working on. All geologists are surely historians as well—we look at ancient occurrences affecting aspects of the earth itself in contrast to Susan's dealings with "old" buildings dating back 50+ years. After all, California is a young state and wooden buildings which pre-date the Gold Rush are rare. My mother was the town historian in Greene, N.Y., which only has a cou-

Here we are in a gondola in Venice...and happily sailing together through life.

ple of buildings constructed in the late 1790s. But historians in the eastern U.S. date buildings back to the earliest years of our country in the 1600s.

All these adventures eventually led me into Sea Ranch politics, and specifically into coastal environmental politics. Everyone on the coast claims to be an environmentalist. Soon I became a co-founder and president of Friends of the Gualala River, fighting timber harvest plans; president of Matrix of Change, a group which educates area children on the local environment; a member of the Gualala River Watershed Council which deals with watershed issues; a member of the Redwood Coast Watershed Alliance which focuses on coastal issues; and I was also a member of the local Rotary, and on and on.

I took the California State exam to become a Registered California Geologist in 1995 and passed it on my first try, in spite of my ancient geologic training being pre-plate tectonics as well as my lack of real geologic knowledge regarding much of California itself. Since then, I've worked upon request on small projects exclusively in Sonoma and Mendocino Counties which are where both my interest and expertise lie.

A few years ago, I decided to become a writer and document some of the interesting things I've learned about the geology along the Northern California coast. While this is my second book, I'm now trying my hand at fiction for my next one which will probably be branded as a detective book as the two heroines solve crimes, but they also deal with major issues of our time. I'm actually already starting its sequel which will be further out there yet. The mind is a dangerous thing and one should be cautious when it leads you into the realm of fiction!

◆ CHAPTER 27 ◆   Reflections on "Big Oil"

Back in the 1960s, when I first went to work for "Big Oil," a couple of us debated with each other about the value of the big international oil conglomerates. Were these ginormous and increasingly geopolitical corporations inherently good or evil—or possibly just amoral? Maybe we were just trying to assuage our conscience in attempting to justify the fact that we worked for one of them!

One thought we had was that these huge international companies were good and that they would suppress wars and uprisings, in order to protect their assets. However, in retrospect, it might appear the opposite was true. Too much oil on the market drives the price down. Possibly a war here or there then provides an advantage for oil from an adjacent country. After the war is over, Big Oil can repair the wells for a sizable fee paid by the government of a given country. It is then back to business as usual. OPEC does a poor job of controlling the price as its individual member countries seem to always cheat on their agreed-upon share of the market.

Although discrimination in hiring minority workers was never specifically mentioned herein, the oil patch was truly a collection of WASP-y "good ole' boys." I never met a black or Hispanic person in any of the companies I worked for nor interacted with—and especially never in the field. There were a few women geologists over the years, but very few. Oklahoma has a very large Native American population, but those working in the oil patch were few in number.

In the U.S., most states have enacted laws to regulate how these companies treat our environment. In third world countries, "Big Oil" has not done a very good job of protecting the environment at all. They have commonly flared the natural gas (for lack of market), rather than re-inject it back into the reservoir to maintain the reservoir's pressure. They cut roads haphazardly, lay pipelines, and leave junk everywhere with apparently little supervision or concern for the damage left in their wake. Their philosophy seems to be: get the big oil production and Get Out—and leave the lesser production and cleanup to someone else!

The same practice is still going on in many areas including here in the U.S. The new production in North Dakota—a state with little regulation on how oil is produced, transported, etc.—illustrates these problems. More casing needs to be set to protect fresh water supplies, and control where the frack goes. Trainloads of tanker cars filled with crude oil have ended up parked in cities and suburban areas. The 2147-mile Keystone pipeline runs from Alberta, Canada to southern Illinois, and then 291 miles to Cushing, Oklahoma, and then an additional 487 miles to Port Arthur, Texas. It was rushed through the permit process in four states. The Dakota Access pipeline is now operating the 1172 miles from North Dakota to southern Illinois. It met with a gigantic protest led by Native Americans who shut down its construction for a time. These pipelines move over a half million barrels of oil per day each and President Trump is supporting both of these pipelines.

The future of oil & gas

Natural gas will be here for a long time into the future. It is cleaner to burn than coal or wood, and safer than nuclear energy. Reserves are abundant. The infrastructure is already in place to support its use in many of our cities nationwide, and gas is cheaper than heating oil or electricity. Wood-burning fireplaces have been eliminated or will be soon in many cites because of the pollution they generate.

Hopefully, in the near future, we can reduce our usage of oil. Green energy is slowly becoming accepted and alternative energy sources are being developed. Automobiles are being powered by electricity, hydrogen, and bio-fuels. Hybrids are becoming very popular and reducing pollution, especially in heavily populated areas and during peak commute times.

Solar, wind, biomass, and geothermal are producing more and more generation of electricity. Hydropower is still an important source for electricity in some areas with existing dams. However, environmentalists are opposed to dams blocking our streams, and have been instrumental in having some dams removed. I do not believe many more existing dams will be removed in the near future, or that new dams will be constructed.

If nuclear waste can be disposed of in a safe manner, then nuclear plants may again be an accepted source of energy—hopefully we recognize by now though that locating a nuclear plant in an area of repeated seismic activity isn't a wise choice.

The big oil companies do not intend to slip away into the dustbin of history though, especially as petroleum has many uses besides just powering our automobiles including in the production of such unexpected things as CDs, deodorant, guitar strings, and even denture adhesive. The chemical industry also uses oil as a major resource. Textiles, medicines, and plastics all need oil, including for things like vitamin capsules.

Additionally, the major industry players have research labs actively working on alternative energy sources. Several have developed geothermal divisions which use drilling, casing, and pipelines to deliver the steam and heat to their electric generation plants. These companies will also continue to exist as we will surely always need some oil.

In the future, we will find some ways to dispose of carbon dioxide, which is forming a blanket and contributing to the rise in temperatures globally. Carbon dioxide has been used by the oil industry to re-pressure oil reservoirs for secondary recovery of oil. Rising sea levels and the acidization of sea water in the oceans is killing off corals and coral reefs. Corals and shelled animals trap carbon dioxide in their shells. Maybe CO_2 could be used somehow to save the coral reefs.

Ultimately, the future energy sources utilized will be defined by the consumers — as well as by the declining reserves of coal, oil, and natural gas. Until recently, the costs of green energy had been higher than these established energy commodities. The coal industry is currently in rapid decline from the competition by low-priced natural gas. Subsidies have allowed solar and wind energy to vigorously compete in the energy market too. This has resulted in newer technologies and a reduction of price, making solar and wind competitive. In third world countries, electricity is often available only part of the time. If a poor family can buy or acquire, a solar panel or a small wind turbine as a gift, they will then have electricity with no future cost to them (during daytime hours at least and/or when the wind is blowing) Their current electricity is erratic and costly, so producing their own electricity would provide a significant economic advantage. Put a little larger solar array on a building for multi-families and, with a rechargeable lithium battery, they can have 24-hour electric service.

Gas and oil from shale

Most scientists believe that oil and gas are born in shale deposits. Shale is mud that has been deposited in quiet water, much of it in deep ocean basins. Organic material rains down to the ocean floor and is buried in the mud. As layers build up over time, the shale is squeezed from the overlying weight of sediments above, and oil and natural gas are generated. It may migrate to overlying beds and be trapped in sandstones, limestones, or other rocks. Much of the organic material remains in the shale. For years, as we drilled through shales, we knew there was oil and gas there as our mudlogger's equipment always displayed gas kicks.

Shale formations occur in all the major basins in the world. The beds vary from a few feet in thickness to hundreds of feet in thickness. In the continental U.S., there are large shale formations in more than 20 states, plus Alaska—and more large shale deposits yet are located in Canada.

The development of the technology to get oil and gas out of these shale deposits allows us to count these as large reservoirs. They probably contain more oil and gas than has been produced from all the oil and gas reservoirs developed to date in the entire world. So now oil shale has negated all the predictions of the past that we were running out of oil and gas!

The big oil companies have all done research on green energy, but have never spent the kind of research dollars needed to significantly move us away from oil and gas. They have played with extraction techniques from shale for many years. Horizontal wells and the super-sized fracture treatments have accomplished recovery at an economic price.

Pricing is still the controlling factor for this all-important commodity which, again, is driven by the ebb and flow of supply and demand. The oil and gas easily found onshore has already been explored and developed (except for shale oil). Big Oil had moved offshore for many years and developed many oil and gas fields. Although drilling and production platforms are expensive, the reserves have been large. Yet many of the offshore basins are still unexplored.

OPEC controlled the price for many years and set production quotas along with their pricing. After the Soviet Union broke up, the Russians have developed their oil fields and flooded the European market with oil and gas, thus breaking the hold of OPEC. The middle East still has tremendous reserves of oil and gas. The Saudis can sell their government-owned oil for as little as $10/barrel and still make a profit.

Oil prices have dropped from $100 per barrel to around $60+ per barrel currently. US oil production from shale oil has driven the price down at one point to $40 per barrel, but it now seems to have stabilized in the $60+ per barrel range.

The current world usage is around 93 million barrels of oil per day and is still slightly increasing. The developed countries usages are slowly declining, but India, SE Asia, and China are rapidly growing—and with their ginormous emerging middle classes, so too is their appetite for energy increasing exponentially. The U.S. has nearly doubled our production from 5.6 million barrels of oil per day in 2010, to 9.3 million barrels per day in 2015 and that output is still growing as well. The U.S. growth is primarily being derived from shale oil production in West Texas and North Dakota, however shale plays are occurring in several other states.

Petroleum geologists, like myself are now obsolete and practically irrelevant for shale oil production—in fact, in 1965, I predicted that geophysics would make geologists obsolete. Certainly, for offshore seismic exploration, geologists have taken a back seat. These days, producing shale oil depends upon technology, with little or no input from geologists.

Drilling for shale oil and gas

Pick a shale bed to explore in any of 20 states in the U.S. Have any clerk check the well records around where you want to drill. What is the depth of the shale bed, and how thick is it? Have a drilling engineer design a drilling and well completion program for that depth. You will need a land plat map showing where the well is to be located and the paths of the horizontal legs. You don't need a geologic report or well cross-sections, and you don't need a wellsite geologist. No one needs to look at the samples and probably you don't need to even obtain them.

A modern drilling rig will drill the well. It is highly automated and the drilling may be almost completely drilled and monitored from a distant location by satellite. Pipe is set over the vertical section of the well. The horizontal legs of the well are three or more in number and controlled by a computer technician back in a windowless office staring at computer monitors. Electric logging tools are attached to the drill bit. A simple gamma ray log will tell if the leg is being drilled in the shale bed. If the drill bit wanders out of the shale bed, the drilling can be adjusted to bring the bit back into the formation.

Once the number of legs have been drilled and cased with slotted liners, the well is ready for fracking. The fracks are fresh water fracks with tons of sand to prop open the fractures created. Various chemicals may be added to facilitate the frack and prevent the swelling of clays in the shale formation. The fracks are in stages and may total up to three to five million gallons of frack water. The frack is accomplished by huge frack trucks with pumps producing very high pressures.

The fracks can be monitored to see where the fracks are going, or are possibly being diverted along faults or into old wellbores. Seismic geophones are placed on the ground surface above the area that is fracked. The fracturing causes mini-waves which can be picked up by the geophones. If all the frack suddenly is channeled, say into a fault or natural fracture, a sleeve may be closed and the frack put in a different part of the leg.

The returning frack water is separated from oil and gas and often disposed of by pumping it into shallow porous formations. This practice is currently causing earthquakes in northeast Oklahoma.

These wells cost several million dollars each to complete. Initial potential rates can be very high—thousands of barrels per day rate. If the producing rate stabilizes at, say, 1000 bbls/day, then the well will make over $1M per month and can pay out in less than a year.

Pricing history

The price of this key commodity has always varied based upon supply and demand. I joined the industry before OPEC nationalized their oil and set the prices artificially. American oil companies had found and developed most of the OPEC oil. Obviously, the oil companies did not get the U.S. to go to war and take over the oil fields. More recently, the Bush Administration thought we could possibly take over the oil from Iraq. Libya had undeveloped oil some U.S. companies wanted. In spite of the revolutions in Iraq, Libya, Egypt, etc., none of these oilfields have been taken over by the big multi-national oil companies.

The oil companies have instead simply settled for the oil supply, as they make their real profit from marketing the end product.

OPEC set quotas on production for their members, but there was always lots of cheating afoot. They did certainly have a big say in the price of crude oil on the world market. Before OPEC, oil produced domestically sold for $4/barrel. Once OPEC came on the scene, they caused the price to instantly jump to around $20/barrel.

After the breakup of the Soviet Union, Russia became a major oil and gas exporter and constructed pipelines to Europe. They did not join OPEC, and the prices went down.

Iran has large reserves of petroleum and has had a very limited market until the new nuclear treaty was signed and, with it, a relaxation of restrictions. The effect of increased production from Iran on the world market will probably keep the price down for the near future.

The U.S. has decided to become self-sufficient. Increased drilling and high prices sparked the North Dakota boom and the fracking of oil shales in several states. Unless we control the domestic oil price and keep it around $50/barrel, I do not think the domestic oil business can be significantly profitable.

Natural gas prices are low because of abundant domestic supplies. The generation of electricity using natural gas is much cleaner than coal, and ultimately will take over and replace all the coal-generated plants.

1973 — OPEC enacts oil embargo

The Arab-dominated Organization of Petroleum Exporting Countries (OPEC) announces a decision to cut oil exports to the United States and other nations that provided military aid to Israel in the Yom Kippur War of October 1973. According to OPEC, exports were to be reduced by 5 percent every month until Israel evacuated the territories occupied in the Arab-Israeli war of 1967. In December, a full oil embargo was imposed against the United States and several other countries, prompting a serious energy crisis in the United States and other nations dependent on foreign oil.

OPEC was founded in 1960 by Saudi Arabia, Iran, Iraq, Kuwait, and Venezuela with the principle objective of raising the price of oil. Other Arab nations and Third World oil producers joined in the 1960s and early 1970s. For the first decade of its existence, OPEC had little impact on the price of oil, but by the early 1970s an increase in demand and the decline of U.S. oil production gave it more clout.

In October 1973, OPEC ministers were meeting in Vienna when Egypt and Syria (non-OPEC nations) launched a joint attack on Israel. After initial losses in the so-called Yom Kippur War, Israel began beating back the Arab gains with the help of a U.S. airlift of arms and other military assistance from the Netherlands and Denmark. By October 17, the tide had turned decisively against Egypt and Syria, and OPEC decided to use oil price increases as a political weapon against Israel and its allies. Israel, as expected, refused to withdraw from the occupied territories, and the price of oil increased by 70 percent. At OPEC's Tehran conference in December, oil prices were raised another 130 percent, and a total oil embargo was imposed on the United States, the Netherlands, and Denmark. Eventually, the price of oil quadrupled, causing a major energy crisis in the United States and Europe that included price gouging, gas shortages, and rationing.

In March 1974, the embargo against the United States was lifted after U.S. Secretary of State Henry Kissinger succeeded in negotiating a military disengagement agreement between Syria and Israel. Oil prices, however, remained considerably higher than their mid-1973 level. OPEC cut production several more times in the 1970s, and by 1980 the price of crude oil was 10 times what it had been in 1973. By the early 1980s, however, the influence of OPEC on world oil prices began to decline; Western nations were successfully exploiting alternate sources of energy such as coal and nuclear power, and large, new oil fields had been tapped in the United States and other non-OPEC oil-producing nations. **Source:** history.com (A&E Networks)

Those of us on the East Coast as well as those of us on the West Coast often have difficulty comprehending how different things are in the Heartland. We joke about the "flyover states" and discount or dismiss people from Oklahoma and Texas — often deriding them as we think they talk funny and may appear to be uneducated. These are often God-fearing people with deep ties to the land, ties which are no doubt strengthened by the fact they often span numerous generations — ties not always understood or appreciated by those of us who live in crowded cities on the right and left coasts. There are surely pervasive cultural differences in terms of traditional vs. progressive values in these divergent regions, and one of the more notable differences includes Midwesterners continuing to embrace the oil and natural gas industry, in spite of the growing evidence of the industry's threat to our planet.

We wonder about these ties of theirs as much of the land in the country's midsection appears to us as marginal, hot and dry, with a few small short trees, or no trees whatsoever in areas. The winds blow endlessly there, and you can't open your mouth or you'll find grit in your teeth.

Oklahoma was indeed marginal land when President Jackson sent the east coast Indians from Florida and the eastern states. The excuse was that the country had prime land in Florida and adjacent states held by Indians and they were preventing people of European descent from their expansion efforts into those regions. As land ran out in the east, the Oklahoma Panhandle was opened for settlement. The "Sooners" were individuals who left before the gun was fired to open the land race. Why did they want it so badly?

There were no trees to build houses with, so they cut sod and created mud bricks. The prairie grass had deep roots and made a good sod brick which resisted the wind and the rain. Partial dugouts and sod houses made for effective insulation from the heat and cold.

In due course, they cleared the land of the indigenous prairie grass — which was also holding the soil — and deeply plowed the earth, as was the standard practice in the East as well as in Europe. A few drought years in a row blew away much of the topsoil and prevented crops from being grown or harvested. Many of the Okies then migrated west and many ended up in California. Contrast the culture and attitudes of Californians, even today, of those people living in the Central Valley with those of us living on the coasts.

Living in the Heartland, people are close to the land and they're tied to it. They depend on wheat crops or raising beef cattle as their primary agricultural activities. Droughts are commonplace and, in a single year or two, a farmer's income can be completely lost as the crops fail and the cattle die. If you live there, you are subject to the whims of Nature, and you accept this as the natural way — God's way. People pray for rain and a better year. The cycle changes and they believe their prayers helped accomplish it.

On the East and West Coasts, people are more optimistic and less accepting of the whims of the natural world. They expect the economy to continue to grow and their incomes to rise, but a big snowstorm in New York City will shut the metropolis down — the kids will love it, but the adults will bitch about the government not cleaning up the white stuff clogging their streets fast enough. The crowding of the East and West Coasts makes many of us more interested in preserving nature, and regulating development as well as industry. We want to recycle and reuse materials and decrease the sizes of our landfills. We want to restrict the cities in size and preserve green belts in between cities. We look to government to accomplish these tasks and governmental regulation to save us from problems.

In Oklahoma and Texas, there is lots of land. The cities sprawl over large regions, but are fewer in number than on the coasts. Hydrocarbons — natural gas and oil — provide most of the energy for electricity here. They don't recycle to any large extent with the excuse being, "Think of the cost of sending ships of cardboard, plastic, paper, and glass to China and South Korea. We are probably using more energy to send it than we're saving in the recycling." There is some truth in this statement of course. If we really want to cut down on greenhouse gases, then we should have factories here in the U.S. to re-process these materials.

But if they love the land, then why do they allow or even want oil and gas wells drilled on their property? Of course, they like the money. But they also don't seem to believe that a small oil spill will do any lasting harm to their land. After all, oil is supposed to be organic, isn't it? I have seen areas that had oil spills and drilling mud scattered over them, and the next year the crops were twice the yield or greener yet in appearance. Although, physical appearances aside, they don't seem to worry about the trace chemicals the wheat or the cows may have picked up!

In mentioning T. Boone Pickens again, his early vision was to acquire more and more natural gas. He bought and merged gas companies and made his fortune off natural gas. But then his focus switched to wind energy, and he's developed or promoted the construction of many wind turbines. Oklahoma is now reported to have 25% of its electricity generated by wind. (And they hardly believe in green energy.)

Attitudes change slowly in the Midwest. We on the East and West Coasts spew out a lot of environmental rhetoric, we beat our chests, and want to restrict and change the world. In the Midwest, when problems develop from lax or a lack of regulation, they fix the problem. The people don't want more government regulation and only look to these institutions when they cannot resolve the problems among themselves. They slowly change the laws or adapt them to their specific situations. Maybe someday all the bars will be smoke free. Maybe they will even recycle. I believe they will slowly go green as it makes economic as well as environmental sense. At some point in the future, they may quit horizontal drilling with large frack jobs. The disposal of the frack waters is causing earthquakes and the people are finally rebelling. They could simply fix the problem by reusing the frack water and not pumping it into the wrong places. It takes more time for these things to be accomplished. However, the coastal practices of massive campaigns and lawsuits, which take years to settle is not their way (look at the Exxon Valdez spill which is still not settled). One wonders which way is best or more efficient?

So, when you fly over these states, consider that their attitudes are not going to change [much], or anytime soon. The oil and gas business will continue well into the future. We are not going to quit eating beef, (although beef consumption is declining) or consuming wheat. The country needs these products as well as the energy generated by using oil and gas. The farmers raising the crops and cattle forget the methane generated by the cattle, ignore the pesticides spread by crop dusters, and don't even consider the gene spliced grains, because most of the population does not see the effects of these actions. The winds rapidly disperse particulates, smoke, and carbon dioxide—and therefore are not considered to be a problem locally.

There are no oceans in the Midwest, so they don't have to worry about sea level rise, so some sentiments from these folks in the middle of the country often include: "Those dumb people should not have built in those low places. Who cares if Malibu floods? We don't like those Hollywood liberals anyway. The climate is cyclical, and we can't control it. There have always been hurricanes and tornados. Are these now more widespread or violent than they used to be? Who knows? Remember the floods of the late 1920s? How about the blizzard of 1888?"

When you live out in the countryside and cities are more than 100 miles apart, it's difficult to think about overpopulation and the global pollution of our atmosphere. These folks understand there are billions of people in China, India, and Southeast Asia, and that they have some very intense problems. But they're just as likely to then think they need to clean up their own problems. Densely populated areas encourage people to desire a reduction of their footprint. Vast open spaces do not seem to have that effect.

Many of the people in the Midwest believe that the biggest problems in rural America are the loss of the family farm to giant agricultural conglomerates, and the control of the crops that they raise.

In looking back on my life as well as those of my parent's and grandparent's from this elevated cultural position we've achieved in the 21st century, I'm in awe that I/we really looked at things so differently "back in the day." I wonder, how did we get here? I can barely return to those earlier times in my mind. Especially as the social and cultural gap between men and women was so great. Men had their place in the sun, were the family breadwinners, and made the decisions for the family. The place for the women was to raise the children and maintain the home.

World War II was the first real change in that women entered the work force in large numbers as the men were away at war. After the war, women were expected to return to the home, but many did not—some out of necessity as their husbands had been killed in battle. However some found they enjoyed the liberation of earning their own money. Soon banks and credit organizations recognized two breadwinners in the family. The Women's Movement began to free females from the inequalities of life and work they'd endured for generations. Men actually began to see women as equals with equal rights.

Throughout most of my life, I have been trapped in the "old ways." For much of my life, I brand myself as "a male chauvinistic pig." I made the decisions for my family(s) and felt it "my right"—I wanted the education I felt Kay had denied me. Obviously, having children was 50% my doing. My fault was a lack of communication with her and the failure of us jointly making a commitment on both of our parts to achieve my goal. It needed to have been our goal. I made the decision to leave the "Big Oil company" — again with little concurrence from Kay at the time.

Ann was different and professed to be a liberated woman. However, she did not support NOW (National Organization for Women) or get excited about women's issues. I became an active supporter of NOW at the time — but, again, I excluded Ann from most of my business decisions. We worked together day in and day out but she deferred to me for most business decisions. Her daddy was the patriarch of their family and made decisions for everyone in the family — no matter what their age. Ann's brother, Boyd, filed for his first divorce the day after daddy died. Earl was against divorce and had forbade Boyd from getting one, regardless of how miserable he claimed he was.

When Susan and I got together in the early 90s, she had just come out of a very controlling marriage. From the beginning, we decided we would operate on equal footing with one another and make decisions together about our lives. We also got involved in each other's businesses, and made decisions jointly about that arena as well. As I would accompany her on fact-finding trips relating to her work as a successful author of historic studies, she would in turn help me deal with contentious partners. I overrode a couple of recommendations of hers (falling back into my old ways) — those actions on my part turned out to be great mistakes. I guess I'm just a slow learner!

There were other things happening at the time which helped us change our personal beliefs on the heels of the equal rights movement from decades earlier: the environmental movement, the gay rights movement, animal rights movement, etc. As a society, we fought "Big Tobacco" and won. Green energy began to develop. Pollution, global warming, sea level rise, drug abuse, the nuclear threat, and on and on have each made demands on our attention.

The quest for energy has given us both the good and the bad. Modern conveniences and a decrease in physical work have provided us with time and personal energy to pursue other things. We've had time to reflect on the directions that society and the world are travelling. Global communication connects us all to what is happening. Everyone now has an opinion on everything! There's no gender or race barriers in this communication or in these movements.

The mass murders, civil wars, suicide bombers, para-military hate groups are all just reactions to the social change which is engulfing our culture here in the States. Those of us who have changed and evolved in our philosophies over our lifetime are the lucky ones. In many parts of the world, it will take a couple more generations for this same degree of changes to occur.

Big international companies, including Big Oil, may take a couple more generations to really adopt a changed attitude. Now they're focused on profit for their stockholders. They claim this is the bottom line, but is it really? Profit today may cause serious environmental problems which will affect profits in the future. For example, the flaring of natural gas from oil wells can be seen from space in North Dakota and Saudi Arabia. If they do not have a current market for this gas, why do they not reinject it into the oil reservoir? In the future, we will need this gas, and if we restrain from burning it in flares, it will reduce particulates, carbon dioxide, and heat in our atmosphere.

So, in actuality, the bottom line is more than just today's stock price and the upcoming quarterly dividend...yes?

◆ CHAPTER 29 ◆   Some final thoughts...

In less than 150 years, humans have progressed from the horse & buggy age, developed gasoline-powered engines, invented independent transportation for individuals as well as moving commodities, grown our cities from short squalid structures to those reaching upward toward the heavens. It all revolves around energy: heretofore coal, oil, and natural gas. Giant oil companies, (predominately American-owned) have lead the search worldwide for hydrocarbons. The study of geology may have developed in Europe and the United Kingdom, but it was grown here in American universities which trained geologists who have developed the natural resources of our planet.

Mining and petroleum geologists travelled to the far reaches of the world. We looked at rocks and fossils, made maps and wrote reports, and uncovered the building blocks and energy needed to fuel our industrial civilization. Other sciences were used to further the search. The masses of data collected required computers to analyze the information gathered. Most geologists now spend their time at computers in offices far removed from the natural landscape.

Geology gave way to technologies which continue to develop and to mine low-grade mineral deposits, and to wring oil and gas out of tight, natural rock formations. With the new technologies developed to deal with shale oil and gas, a geologic knowledge of the rock formation and its geologic setting is barely needed, if indeed needed at all. Big Oil companies can now operate with smaller staffs, drill fewer wells, and yet effectively produce more oil and gas yield. They have taken on the philosophy that green energy may be good for the environment, but their source of oil and gas is assured for maybe hundreds of years into the future. The world will continue to rely upon oil and gas for transportation, or the generation of electricity for transportation. The power grid can be easily regulated by using natural gas, when the wind is not blowing and the sun is not shining.

Geologists like myself, can now go back out to the field and study nature. We can map pollution and erosion caused by human activities in our search for minerals. We can advise and educate people about geologic hazards, pollution, and methods of cleanup and preservation of our natural environment.

So I've returned to my geological and environmentalist roots, especially as the future is in alternative green energy. Solar and wind are the new choices in generating power. Hydrogen is also promising, as are biofuels made of algae, etc. Wave energy and tidal movement could provide a never-ending source of electricity, but are fought by many environmentalists.

Geothermal energy is very promising, but needs some innovations to reduce the costs of geothermal wells and power plants. I have a great prospect (another wildcat deal) over near the Geysers Steam Field in northern California. Several years ago, I paid the Bureau of Land Management (BLM) a fee to put up for competitive bidding a couple thousand acres, which I feel have lots of potential. I finally got a call from the BLM a few months ago telling me they were working on an environmental report. Ah, the government bureaucracy moves so slow-w-l-y! Geothermal exploration continues in many parts of the world. Iceland is completely dependent on heat energy for example. Drilling geothermal wells is expensive though, while wind and solar costs for electricity generation have become very affordable. The Geysers field is being re-charged with waste water from Santa Rosa, but the injection of the wastewater is creating small earthquakes, which may develop into an environmental issue in the near future.

The major oil companies all are looking or experimenting with green energy. BP advertises their name means "Beyond Petroleum." However, in the most recent election cycle, they kept running these ads with "little ordinary people" saying they're energy voters. What does that mean? More coal, more drilling like in North Dakota, and more fracking? I consider myself an energy voter—and I now vote for green energy!

All of the above are big projects or quickly grow into a big project. Alternatively, we need small, less environmentally-intrusive sources of power—like biomass, or heat pumps—something simple. With solar panels on our roof, we have become energy self-sufficient—enough to power our own home plus an electric car. Everyone we know is recycling part of their garbage (it's even happening in Oklahoma!). We could do more composting of the organic waste we generate—which might even lead to a local biomass project.

Become a "steward of the land" and think globally, think regionally, think locally—and then think smaller and more individually still. Save a tree at a time, save a tributary, save the watershed, save a beach, then save the ocean…and, collectively, we can protect and ultimately safeguard the future of our fragile planet.

After all, shouldn't our goal be to leave our planet better than we found it—for our grandchildren? Or, as the Native Americans say, for the seventh generation hence…

Reflections of mine...

Life's journey is a marvelous adventure.
And around every corner, there's another new fork in the road.
The question that always faced me was
"Which fork shall I choose?"
Suppose I had taken the fork offering the second summer
on the Juneau Icefield?
Would I never have then worked for '"Big Oil?"
Would the Leprechauns never have
bitten me with that "wildcat bug?"
How did I discover the fork which led me
to The Sea Ranch and
Susan, the love of my life?
Who placed these forks in my personal road?
How did I know the right ones to take?
After all, which one *was* the right fork?
Or the wrong fork?
Do we really have "free will?"
Or is it God, or our genes which
determine our path?
I guess I won't worry about it.
Once I had crossed the Rubicon, the die was cast!
Upon reflection, my life was always busy...and it was fun.
So...do you want to drill a well? I know where the black gold is *still* hidden!

Questions? Comments? Or otherwise want to get in touch? **Please visit:**

www.RiverBeachPress.com

Or drop a note:

Thomas E. Cochrane
PO Box 358
The Sea Ranch, CA 95497

If you enjoyed **Tornados, Rattlesnakes & Oil**

please leave a review on **amazon**

And **THANKS** *for any other sharing you might do on this book's behalf...*

Meet the author...

Thomas English Cochrane is a California Professional Geologist (#6124). Born and raised in Greene, New York, he wandered and roamed the area's glacially-formed hills and valleys while growing up including the Catskills, the Adirondacks, and the Appalachian mountains. These youthful explorations led to him obtaining an under-graduate degree in Geology at the State University of New York (SUNY), Binghamton. He subsequently did graduate work in Education at Colgate University, graduate work in Geology at Indiana University, and a study in field geology at Miami of Ohio's Geology Field Camp in Wyoming.

In the early part of his career, he taught science, mathematics, earth science, and physics in his hometown at Greene Central School, later becoming the chairman of the Science Department and acting Chairman of the Mathematics Department. He then received a National Science Foundation Grant to study Glaciology for a summer on the Juneau Icefield in southeast Alaska.

Mr. Cochrane began his 24-year career in the oil and gas business in 1964 in Oklahoma with Pan American Petroleum. He left Pan Am in 1968 and spent the next 20 years working as a consulting petroleum geologist. During this time, he was also editor of the *Shale Shaker*, a geologic publication of the Oklahoma City Geological Society.

In 1988, he moved to The Sea Ranch in Northern California where he'd been spending time for over a decade, and began consulting on geologic hazards and local geology along the coast. He became a California Registered Geologist in 1995 (this designation was recently reclassified as a California Professional Geologist). He is the author of another book, ***Shaping the Sonoma-Mendocino Coast—Exploring the Coastal Geology of Northern California*** (River Beach Press, 2017), and he regularly addresses regional groups and organizations on topics relating to coastal geology. Mr. Cochrane was recently named Director of the Redwood Coast Land Conservancy, a 501(c)3 nonprofit which is a member of the National Land Trust Alliance and California Northern Region Land Trust Council.